Foliage Plant Diseases

Diagnosis and Control

A. R. CHASE

Chase Research Gardens, Inc.
Mt. Aukum, California

APS PRESS

The American Phytopathological Society
St. Paul, Minnesota

The American Phytopathological Society
3340 Pilot Knob Road
St. Paul, MN 55121-2097, USA

Contents

Introduction

Foliage plants have played an integral role in interiorscapes of private and public spaces since the 1920s in the United States. Use of plants in homes began 3,500 years ago when the Sumerians and Egyptians started growing small trees in containers. The ancient Chinese expanded the variety of plants used for this purpose. Development of greenhouses began in Europe about 600 years ago, and the structure attained many of its current characteristics during the 1800s. Information on diseases of foliage plants came hand-in-hand with the increased production that occurred during the 1920s in the United States. In-depth reports of specific diseases were relatively rare until the 1970s, when the foliage plant industry expanded dramatically. The most significant event in the research on foliage plants was the establishment of the University of Florida's Agricultural Research Center in Apopka in 1968. Although many other research centers around the world have contributed to the wealth of information on production of the foliage plants currently available, no other single facility has existed solely to serve the foliage plant industry. Because of the budget constraints across the United States, publicly funded research on ornamental plants of all types has been decreasing since the late 1980s.

Since the 1960s, the number and variety of foliage plants produced commercially has continued to increase annually. More than 500 species of foliage plants are currently produced under a wide variety of both technical and common names (Table 1). Florida accounts for 36% of the producers and 76% of the area under production. California, Hawaii, New York, Ohio, Pennsylvania, and Texas are also significant in terms of number of producers or area under production. A variety of books is available that at least in part contain information on foliage plant diseases (listed at the end of this section). This book was developed to summarize current information on foliage plant diseases, with special emphasis on symptom recognition via color photographs. The cause of each disease is listed along with symptoms, control, and references.

The control sections contain information on chemical disease control that has been tested experimentally. Many times no chemical controls are mentioned, indicating a lack of knowledge regarding a specific disease on that host. Table 2 has been supplied to allow interested readers to find some of the commercially marketed products for each active ingredient. Mention of a product does not imply any warranty or recommendation for use. Similarly, failure to mention a product does not imply criticism of that product. Avoiding disease through cultural management is always recommended, whether or not chemical means are employed. The most important factor in maximizing the benefits of pesticides is that all other control methods are also employed. Using pesticide inappropriately results in poor control, labor and product loss, and sometimes phytotoxicity (Table 3). Read the label on each pesticide you use, and consult local state authorities to be certain that you can legally apply the product (under your conditions) to your crop. Remember: the label is the law!

Symptoms and Diagnosis

The diagnosis of foliage plant diseases through symptomatology alone is not recommended, although a few helpful generalizations can be made. Symptoms caused by infectious and noninfectious agents are often similar. Accurate diagnosis requires evaluation of affected plants for evidence of infectious agents through direct observation and/or culturing (Table 4).

Some of the more typical symptoms of diseases caused by bacteria are angular lesions bordered by leaf veins. These lesions are frequently surrounded by a yellow margin or water-soaked area. Infections caused by *Erwinia* spp. usually result in some tissue disintegration, giving rise to irregularly shaped lesions and mushy rots (often accompanied by a fishy, rotten odor). It is interesting to note that pure cultures of many isolates of *Erwinia* give rise to the same odor and its generation is not dependent upon rotting plant matter. Lesions caused by *Xanthomonas* spp. on foliage plants are commonly dry appearing and corky in texture and can be easily confused with other bacterial or fungal diseases, phytotoxicity (Table 4), or nutrient imbalance. Culture of the bacterium on artificial media is almost always required for an adequate diagnosis.

Fungi cause a wide variety of symptoms on foliage plants, including root rot, stem cankers, and leaf spots. Although these symptoms can be distinctive for certain diseases, they are easily confused with similar symptoms caused by other plant pathogens, such as bacteria and nematodes, or abiotic problems, such as minor element toxicity. Some fungal diseases (e.g., downy and powdery

TABLE 1. Nomenclature of common foliage plants

Family
  Scientific name
	Common name	Family
		  Scientific name
			Common name

Acanthaceae		Cactaceae (*continued*)	
Aphelandra squarrosa	Zebra plant	*Schlumbergera bridgesii*	Christmas cactus
Fittonia Verschaffeltii	Nerve plant	*Schlumbergera truncata*	Thanksgiving cactus
Agavaceae		Commelinaceae	
Cordyline terminalis	Ti plant	*Rhoeo discolor*	Moses in a basket
Dracaena spp.	Dracaena	*Rhoeo spathacea*	Oyster plant
Dracaena Massangeana	Corn plant	*Siderasis fuscata*	Siderasis
Dracaena Sanderana	Belgian evergreen	*Tradescantia* spp.	Spiderwort
Sansevieria trifasciata	Snake plant	*Zebrina* spp.	Wandering jew, inch plant
Yucca aloifolia	Spanish bayonet	Crassulaceae	
Yucca elephantipes	Soft-tip yucca, spineless yucca	*Crassula argentea*	Jade plant
Araceae		*Sedum* spp.	Stonecrop
Aglaonema spp.	Aglaonema	*Sempervivum* spp.	Hen and chicks
Aglaonema commutatum	Silver Queen and others	Droseraceae	
Aglaonema modestum	Chinese evergreen	*Dionaea muscipula*	Venus's fly-trap
Anthurium spp.	Anthurium, tailflower	Euphorbiaceae	
Caladium × hortulanum	Fancy-leaved caladium	*Codiaeum variegatum*	Croton
Dieffenbachia spp.	Dumb cane, dieffenbachia	*Euphorbia Milii*	Crown-of-thorns
Dieffenbachia maculata	Perfection and others	Gesneriaceae	
Epipremnum aureum	Pothos, devil's ivy	*Aeschynanthus pulcher*	Lipstick vine
Monstera deliciosa	Swiss-cheese plant	*Columnea* spp.	Goldfish plant and others
Philodendron spp.	Philodendron	*Episcia cupreata*	Flame violet
Philodendron scandens		*Saintpaulia ionantha*	African violet
subsp. *oxycardium*	Heart-leaf philodendron	Lamiaceae	
Philodendron selloum	Selloum	*Coleus × hybridus*	Flame nettle and others
Spathiphyllum spp.	Peace lily, spathe flower	*Plectranthus australis*	Swedish ivy
Syngonium podophyllum	Nephthytis	Leeaceae	
Xanthosoma sagittifolium	Xanthosoma	*Leea coccinea*	West Indian holly
Araliaceae		Liliaceae	
Brassaia actinophylla	Australian umbrella tree, schefflera	*Aloe variegata*	Aloe
		Asparagus spp.	Asparagus fern
Dizygotheca elegantissima	False aralia	*Chlorophytum comosum*	Spider plant
x-*Fatshedera lizei*	Tree ivy	*Hawarthia* spp.	Wart plant
Fatsia japonica	Japanese fatsia	Marantaceae	
Hedera helix	English ivy	*Calathea* spp.	Calathea, rattlesnake plant, and others
Polyscias Balfouriana	Balfour aralia		
Polyscias fruticosa	Ming aralia and others	*Ctenanthe amabilis*	Dragon tracks
Schefflera arboricola	Dwarf or Hawaiian schefflera	*Maranta erythroneura*	Red maranta
Tupidanthus calyptratus	Tupidanthus	*Maranta leuconeura*	Prayer plant, rabbit tracks, green maranta
Araucariaceae			
Araucaria heterophylla	Norfolk Island pine	Moraceae	
Arecaceae		*Ficus benjamina*	Weeping fig
Caryota mitis	Fishtail palm	*Ficus elastica*	India rubber tree
Chamaedorea elegans	Parlor palm	*Ficus lyrata*	Fiddle-leaf fig
Chamaedorea erumpens	Florida hybrid palm, bamboo palm	*Ficus pumila*	Creeping fig
Chrysalidocarpus lutescens	Areca palm	Piperaceae	
Gronophyllum Ramsayi	Kentia palm	*Peperomia* spp.	Radiator plant
Phoenix Roebelenii	Pygmy date palm	*Peperomia argyreia*	Watermelon peperomia
Rhapis excelsa	Lady finger palm	*Peperomia caperata*	Emerald ripple
Washingtonia robusta	Mexican fan palm	*Peperomia obtusifolia*	Baby rubber plant
Asclepiadaceae		Polypodiaceae	
Hoya carnosa	Wax plant	*Asplenium nidus*	Bird's-nest fern
Asteraceae		*Cyrtomium* spp.	Holly ferns
Senecio sp.	String of pearls	*Davallia fejeensis*	Rabbit's-foot fern
Aizoaceae		*Dryopteris* spp.	Shield or wood ferns
Faucaria spp.	Tiger-jaws	*Nephrolepis exaltata*	Boston fern
Begoniaceae		*Platycerium* spp.	Staghorn ferns
Begonia × rex-cultorum	Rex begonia	*Pteris* spp.	Table ferns
Bignoniaceae		Saxifragaceae	
Radermachera sinica	Radar plant, China Doll	*Tolmiea menziesii*	Piggyback plant
Bromeliaceae		Urticaceae	
Aechmea fasciata	Urn plant	*Pellionia Daveauana*	Trailing watermelon begonia
Cryptanthus spp.	Earth star	*Pellionia pulchra*	Satin pellionia, rainbow vine
Cactaceae		*Pilea Cadierei*	Aluminum plant
Cereus peruvianus	Apple cactus, Peruvian apple	*Pilea microphylla*	Artillery plant
Ferocactus spp.	Barrel cactus	Vitaceae	
Opuntia spp.	Prickly pear cactus	*Cissus antarctica*	Kangaroo vine
Rhipsalidopsis gaertneri	Easter cactus	*Cissus rhombifolia*	Grape ivy

TABLE 2. Trade names and common names of some bactericides and fungicides [a]

Active ingredient(s)	Trade name(s)	Active ingredient(s)	Trade name(s)
Captan	Captan, Captec, Orthocide	Mancozeb	Dithane, Fore
Chlorothalonil	Bravo, Daconil 2787, Exotherm Termil, Thalonil	Metalaxyl	Subdue
		Myclobutanil	Systhane
Chlorothalonil + thiophanate methyl	ConSyst	Oxycarboxin	Plantvax
Copper oxychloride	C.O.C.	PCNB	Engage, PCNB, Terraclor, Turfcide
Copper sulfate	Cuproxat		
Copper sulfate (basic)	CP-Basic Copper	Propamocarb	Banol
Copper sulfate pentahydrate	Phyton 27	Streptomycin sulfate	Agri-mycin
Cupric hydroxide	Blue Shield, Champ, Champion, Kocide	Thiophanate methyl	Cleary's 3336, Domain, Fungo, Fungo Flo, SysTec 1998
Dinocap	Karathane WD	Thiophanate methyl + etridiazole	Banrot
Etridiazole	Terrazole, Truban	Thiophanate methyl + mancozeb	Duosan, Zyban
Ferbam	Carbamate	Triadimefon	Bayleton, Strike
Folpet	Folpet, Ortho Phaltan	Triflumizole	Terraguard
Fosetyl aluminum	Aliette	Triforine	Funginex, Triforine
Iprodione	Chipco 26019	Vinclozolin	Ornalin
		Zineb	Zineb

[a] This table is given as a reference for common and trade names of some chemicals used on foliage plants only. Inclusion of a product in this table is not meant as a recommendation of that product. Omission of a product is an oversight and does not indicate any comment on a product's safety or efficacy.

TABLE 3. Phytotoxicity of bactericides and fungicides on some foliage plants [a]

Plant	Agri-brom	Agri-Mycin	Ali-ette	Ban-rot	Cap-tan	Carba-mate	Chipco 26019	Cleary's 3336	Daco-nil	Do-main	Kocide 101	Man-zate	Orna-lin	Sub-due	Terra-clor	Terra-guard	Tru-ban	Zineb	Zyban
Aechmea									DE	S		S				S			
Aeschynanthus	DF								S			S							S
Aglaonema			S	S	S		S		S		S	S	S	S			S		
Aphelandra			S				S		S			S	B	S					
Araucaria				S			S		S			S	S						
Begonia			C				S		S			S	S	S			S		S
Brassaia	S	G	S	S	S	D	S		D		S	S	S	S		A	S	S	S
Calathea	S						S		S			S	S	S			S		
Chamaedorea		S	S	S	S				S		B	S					S	S	
Chrysalidocarpus				S	A	S	S		S		E	S	S	S	S	S	S	S	
Cissus			S				S		S			S	B						
Codiaeum							A		S			A	D		S				
Cordyline							S		S			S	S						
Dieffenbachia	S	S	S		S		S		S		S	S		S					
Dizygotheca							S		S					S					
Dracaena	C	S		S	S		S	S	S	S	S	S	B			S	S	S	S
Epipremnum	S	S	S	S	S		S	S	S	S	S	S		S	S	S	S		S
Episcia							S		S			S	S						
Fatsia								S	S	S	S	S	S						
Ficus	DE		S	S			S		S		S	S		S			S		
Fittonia		C	S	S			S		S		C	S	S						
Hedera	E	B	S	S			S		E		S	S			A	S	S		
Hoya							S		S			S	S	S			S		
Maranta	S			S			S		S			S	S				S		
Nephrolepis	S	A		S	S	S	S		S	S	S	A	A		S	S	S	S	S
Peperomia		S					S		S			S	S	S	A		S		
Philodendron		S		S	S		S		S		S	S		S			S	S	
Pilea			S				S		S			S	S	S					
Polyscias	S		S						S			S							S
Radermachera												S			S	S			
Saintpaulia	H		S				G						S	S					
Sansevieria				S			S		S										
Schlumbergera	H						S		S			S	S						
Spathiphyllum			S	S	S		S	S	S	S		S	S	S	S	S	S		S
Syngonium	S	S	S		S		S		S		S	S				S		S	
Yucca									S		S	S					S		

[a] A square without a letter indicates that information is not available regarding safety of that compound on that plant. S = safe use. Other letters indicate plant injury, at least under some circumstances or conditions. A = stunting; B = chlorosis; C = marginal burning; D = distortion of new leaves; E = necrotic spots; F = leaf cupping and drop; G = stunting and chlorosis; and H = chlorotic spots on flowers.

TABLE 4. Symptoms and possible causes of problems found on indoor foliage plants

Symptom	Possible cause(s)	Symptom	Possible cause(s)
Leaves		Stems	
Powdery coating on leaves	Powdery mildew fungus	Stem rot at soil line	High soil salinity
Speckling on leaves	Mite or insect feeding		Slow-release fertilizer placed against stems
Young leaves chlorotic	Iron too low		Overwatering
	Poor soil drainage		Poor soil drainage
Older leaves chlorotic	Low nitrogen, magnesium, or potassium		Stem disease caused by bacteria or fungi
	High soil salinity	Stem lesions or cankers	Sun scald
	Overwatering		Mechanical injury
	Poor soil drainage		Stem disease (caused by bacteria, viruses, or fungi)
	Root disease caused by pathogens		
Chlorotic spots on leaves	Cold-water injury	Stem cracks	Mechanical injury
Water-soaked spots on leaves	Temperature extremes		Stem disease (caused by bacteria or fungi)
	Bacterial infections	Stem thin and weak	Fertilizer extremes
Necrotic margins or tips of leaves	Boron or fluoride toxicity		Low light
	High soil salinity	Roots	
	Temperature extremes	Poor root development	High soil salinity
	Drying		Soil temperature extremes
Necrotic spots on leaves	Cold water injury		Plant potted too deeply
	Fertilizer toxicity	Root rots	Root disease caused by pathogens
	Sun scorch		High soil salinity
	Cold injury		Poor soil drainage
	Bacterial, fungal, or viral infections	Entire plant	
Leaves too large	Fertilizer excess	Plant stunted	Root disease caused by pathogens
	Low light		Fertilizer extremes
Leaves too small	Fertilizer deficiency		Low light
	Copper deficiency		Temperature extremes
	High soil salinity		Poor soil drainage
	Root-bound plants		Pot-bound root system
Leaves long and narrow	Low light	General chlorosis	High light intensity
Leaves very thin	Excess nitrogen		High soil salinity
	Low light		High leaf temperature and cold soils
Holes in leaves	Mechanical injury		
	Insect feeding	Wilting	Insufficient soil moisture
Defoliation (leaf drop)	High soil salinity		Low humidity
	Abrupt light reduction		Poor root system
	Chilling injury		
	Drying		
	Poor soil drainage		
	Low soil moisture		

mildews and rusts) are readily diagnosed through microscopic examination alone, while others (e.g., anthracnose, Helminthosporium leaf spot, Rhizoctonia stem rot, and Pythium root rot) can be diagnosed only through culture of the organism on artificial media.

Symptoms of nematode infestations include galling of roots, stunting of both tops and roots, and root lesions. Foliar nematodes live within the leaves of the plant. They produce grayish, water-soaked interveinal areas that eventually yellow and die. Nematodes cannot be cultured on artificial media. They are often detected with sieving techniques and microscopic identification or by direct examination of affected root systems.

Viruses cause symptoms such as mosaics, ring spots, leaf and stem distortions, and stunting. They cannot be cultured on artificial media, and their diagnosis requires specialized techniques such as examination with a light or electron microscope and serology (antibody tests are widely available for the majority of important viral diseases of ornamentals).

Disease Development

Many factors influence the development of plant diseases. Indeed, disease cannot occur unless the following elements are present simultaneously: a susceptible host, a virulent pathogen, and a favorable environment. Some diseases cannot occur without the presence of a vector or carrier agent as well. This is true of certain viral diseases that are vectored by an insect or a nematode. Taking these factors into consideration allows for the development of effective control strategies. Disease control in foliage plants is especially amenable to the inclusion of cultural modifications, such as sanitation or altering irrigation practices or temperature, since many plants are grown in enclosed structures. Additional factors, such as host plant nutrition, plant growth stage, and cultivar, are important considerations in disease control strategies.

References to Diseases of Ornamentals

Chase, A. R. 1987. Compendium of Ornamental Foliage Plant Diseases. American Phytopathological Society, St. Paul, MN. 92 pp.

Chase, A. R., and T. K. Broschat. 1991. Diseases and Disorders of Ornamental Palms. American Phytopathological Society, St. Paul, MN. 56 pp.

Daughtrey, M., and A. R. Chase. 1992. Ball Field Guide to Diseases of Greenhouse Ornamentals. Ball Publishing, Geneva, IL. 218 pp.

Daughtrey, M. L., R. L. Wick, and J. L. Peterson. 1995. Compendium of Flowering Potted Plant Diseases. American Phytopathological Society, St. Paul, MN. 90 pp.

Drees, B. M. 1992. Pest Management Alternative for Commercial Ornamental Plants. Texas Association of Nurserymen. 140 pp.

Fletcher, J. T. 1984. Diseases of Greenhouse Plants. Singapore: Kyodo Shing Loong Printing Industries Pte. Ltd. 351 pp.

Forsberg, J. L. 1975. Diseases of Ornamental Plants. 2nd ed. University of Illinois Press, Urbana. 219 pp.

Henley, R. W. 1983. A Pictorial Atlas of Foliage Plant Problems. Central Chapter Florida Foliage Assn. 40 pp.

Horst, K. R. 1979. Westcott's Plant Disease Handbook. 4th ed. New York: Van Nostrand Reinhold Co. Inc. 803 pp.

Jarvis, W. R. 1992. Managing Diseases in Greenhouse Crops. American Phytopathological Society, St. Paul, MN. 288 pp.

Joiner, J. N. 1981. Foliage Plant Production. New Jersey: Prentice Hall, Inc. 614 pp.

Keim, R., and W. A. Humphrey. 1987. Diagnosing Ornamental Plant Diseases. An Illustrated Handbook. The Regents of the University of California, Division of Agriculture and Natural Resources. 32 pp.

Pirone, P. P. 1978. Diseases and Pests of Ornamental Plants. 5th ed. New York: Ronald Press. 546 pp.

Powell, C. C., and R. K. Lindquist. 1992. Ball Pest & Disease Manual. Disease; Insect and Mite Control on Flower and Foliage Crops. Ball Publishing, Geneva, IL. 30 pp.

Powell, C. C., and R. Rossetti. 1992. The Healthy Indoor Plant. Rosewell Publishing Inc., Columbus, OH. 297 pp.

Steele, J. 1992. Interior Landscape Dictionary. Van Nostrand Reinhold. New York, NY. 149 pp.

Strider, D. L. 1985. Diseases of Floral Crops. Vols. 1 and 2. Praeger Scientific, Westport CT. pp. 638, 579.

Aeschynanthus (Lipstick vine)

Lipstick vines are epiphytic plants native to tropical Asia. They are used primarily in hanging baskets in the interiorscape. Plants are produced under 1,500 to 3,000 ft-c with minimum night temperatures of 60°F (15°C). Soil temperatures below 65°F (18°C) will cause growth to slow down or stop. In the interiorscape, lipstick vines need 100 to 150 ft-c to maintain an attractive appearance but may not flower at this light level. Blooming generally occurs in the spring at light levels above 6,000 ft-c. Diseases of lipstick vines are primarily caused by fungi that attack roots, stems, and foliage. Spider and tarsonemid mites, mealybugs, and scales are common pests.

Botrytis blight

Figure 1

Cause   *Botrytis cinerea*

Signs and symptoms   Botrytis blight usually appears on lower leaves of cuttings in contact with the potting medium. The water-soaked lesion may enlarge rapidly to encompass a large portion of the leaf blade or even the entire cutting. The area turns necrotic and dark brown to black with age. When night temperatures are cool, day temperatures warm, and moisture levels high, the pathogen readily sporulates on both leaves and flowers, covering them with grayish brown, dusty masses of conidia. Cuttings rooted during the winter are especially susceptible to Botrytis blight, since the environment is ideal for the disease and very poor for rapid growth of the plant.

Control   Controlling Botrytis blight of lipstick vine is particularly important during the winter months. Cultural methods that improve foliage drying and reduce moisture condensation on foliage during the night reduce Botrytis blight. Iprodione reduces severity of this disease on lipstick vine.

Corynespora leaf spot

Figure 2

Cause   *Corynespora cassiicola*

Signs and symptoms   Lesions appear first as tiny, sunken areas that are slightly brown. These areas enlarge to about 1 cm in diameter and darken with age. A bright purple or red margin and a chlorotic halo about 1 mm wide are usually present on this host. Leaf abscission is common under optimal conditions for disease expression. Similar symptoms are seen on other gesneriads such as *Nematanthus* and *Columnea* spp. and African violet.

Control   Use the same cultural controls as mentioned for Botrytis blight. Chemical control trials have indicated that both mancozeb and chlorothalonil provided excellent disease control.

Selected References

Chase, A. R. 1982. Corynespora leaf spot of *Aeschynanthus pulcher* and related plants. Plant Dis. 66:739-740.

Chase, A. R., and T. A. Mellich. 1992. 1991 Fungicide Tests for Control of Foliar Diseases Caused by *Botrytis, Corynespora, Fusarium* and Powdery Mildew on Ornamentals. University of Florida, Central Florida Research and Education Center-Apopka, CFREC-A Research Report, RH-92-4.

Myrothecium leaf spot

Figure 3

Cause   *Myrothecium roridum*

Signs and symptoms   Lesions generally appear at the edges, tips, and broken veins of leaves. Necrotic areas are dark brown and initially appear water soaked. Examination of the bottom leaf surface generally reveals sporodochia, which are irregularly shaped and black and have a white fringe of mycelium. Sporodochia form in concentric rings within the necrotic areas.

Control   Use of fungicides when temperatures are between 70 and 85°F (21 and 32°C), minimizing wounding, and fertilizing at recommended levels contribute to minimizing severity of Myrothecium leaf spot of foliage plants. Chlorothalonil and iprodione have been effective for control of *Myrothecium* on other foliage plants.

Selected Reference

Chase, A. R. 1983. Influence of host plant and isolate source on Myrothecium leaf spot on foliage plants. Plant Dis. 67:668-671.

Rhizoctonia aerial blight

Figures 4, 5, and 6

Cause   *Rhizoctonia solani*

Signs and symptoms   Rooted cuttings may be completely covered by a mass of brownish mycelia. Growth of mycelia from the potting medium onto larger plants can escape notice and give the appearance that plants have been infected from an aerial source of inoculum. Close examination, however, generally reveals the presence of mycelia on stems prior to the development of obvious foliar symptoms. This disease is most common during the hottest times of the year when the plant foliage remains wet for long periods or relative humidity is high.

Control   Use pathogen-free cuttings and new pots and potting media, and avoid extremes in soil moisture. Chemical control of diseases caused by *Rhizoctonia* has been investigated on many plants using a variety of fungicides. The fungicide most widely used as a soil drench for control of *Rhizoctonia* diseases is thiophanate methyl. Foliar applications of chlorothalonil have been effective in protecting leaves from *Rhizoctonia* infection on other plants, but remember that the potting medium or soil is a common source of this soilborne plant pathogen.

Aglaonema

Aglaonemas are native to tropical Asia and are usually herbaceous plants less than 1 m tall. They are valued interiorscape plants because of their ability to tolerate low light (75 ft-c) and their striking foliage patterns. The newest cultivars have been chosen for their red or pink petioles. Most aglaonemas are used as small specimen plants or massed in planters. Aglaonemas are usually produced under 1,000 to 2,500 ft-c with a minimum greenhouse temperature of 65°F (18°C). Aglaonemas are subject to a multitude of bacterial, fungal, and viral diseases, which attack leaves, stems, and roots. Scales and mealybugs can cause significant problems on aglaonemas.

Anthracnose

Figure 7

Cause   *Colletotrichum* spp.

Signs and symptoms   Leaf spots are initially tan and water soaked and may have a bright yellow halo. Fruiting bodies of *Colletotrichum* spp. appear in concentric rings of tiny, black specks on the upper surface.

Control   Keep foliage dry, and protect from cold water drips caused by condensation on overhead structures. Many fungicides, such as mancozeb, iprodione, and thiophanate methyl are effective.

Selected Reference

Graham, S. O., and J. W. Strobel. 1958. The incidence of anthracnose fungi on ornamental foliage plants in Washington State greenhouses. Plant Dis. Rep. 42:1294-1296.

Bent tip

Figures 8 and 9

Cause   Unknown

Signs and symptoms   The terminal leaf spike will have a fish-hook appearance, and some older leaves will also have a hook at the terminal. The new leaf tip appears to be obstructed and caught by the succeeding leaf, resulting in the fish-hook appearance. Bent tip, most commonly found in *Aglaonema commutatum* 'Silver Queen,' has been a problem for many years. A new leaf emerges with the tip bent back, usually parallel to the main portion of the leaf. The bent portion can be less than 1 mm to more than 2 cm. If the bent portion is large, it will frequently tear along the connecting tissue as the main leaf expands, causing unsightly damage. High temperatures, high light levels, and reduced moisture will increase the incidence of bent tip.

Control   There are no known control measures at this time, although excessive light and water stress have been observed to increase severity in certain cultivars.

Boron toxicity

Figure 10

Cause   Excessive amounts of boron in fertilizer

Signs and symptoms   Marginal and tip chlorosis is followed by necrosis. Spots are roughly elliptical and can be water soaked and greatly resemble those caused by plant pathogens such as *Erwinia* and *Xanthomonas* spp. Symptoms are also easily confused with those caused by fluoride toxicity.

Control   Never apply a fertilizer that is high in boron. Check the potting media or soil for excessive amounts of boron prior to its use. Some *Aglaonema* cultivars such as Maria are especially sensitive to both boron and fluoride toxicity. If this type of spot appears in an interiorscape, it is more likely caused by nutrient toxicity than by a bacterial pathogen.

Selected Reference

Poole, R. T., and C. A. Conover. 1985. Boron and fluoride toxicity of foliage plants. Agricultural Research and Education Center, Univ. of Florida, AREC-Apopka Research Report, RH-85-19.

Chilling injury

Figure 11

Cause   Air temperatures below 55°F (12°C)

Signs and symptoms   Mainly middle-aged to older (lower) leaves develop gray splotches and become chlorotic; lower leaves may collapse after 3 to 7 days if damage is severe. The cultivar Silver Queen is especially sensitive to cold. Table 5 lists reactions of some aglaonemas to chilling temperatures.

Control   Keep production air temperatures at least 55°F (12°C) for Silver Queen and 45 to 50°F (10°C) for most other cultivars to prevent damage. The damage is permanent, but plants will produce healthy leaves when air temperatures are adequate, unless the shoot tip has been damaged by extreme cold.

TABLE 5. Response of some *Aglaonema* species and cultivars to chilling temperatures during the summer and winter[a]

Aglaonema species and cultivar	Chilling injury (%)	
	Summer	Winter
nitidum 'Curtisii'	68	88
× 'Silver Queen'	57	90
× 'Fransher'	14	59
commutatum 'Treubii'	11	17
tricolor 'Tricolor'	1	0
rotundum	0	0
commutatum 'Pseudobracteatum'	0	0
simplex	0	0
modestum	0	0
pictum 'Tricolor'	0	0

[a] Adapted from Hummel and Henny, 1986.

Selected References

Fooshee, W. C., and D. B. McConnell. 1987. Response of *Aglaonema* 'Silver Queen' to nighttime chilling temperatures. HortScience 22:254-255.

Fooshee, W. C., and D. B. McConnell. 1980. Effect of chilling on subsequent rooting of *Aglaonema* 'Silver Queen' tip cuttings. Proc. Fla. State Hortic. Soc. 93:212-213.

Hummel, R. L., and R. J. Henny. 1986. Variation in sensitivity to chilling injury within the genus *Aglaonema*. HortScience 21:291-293.

Copper deficiency

Figure 12

Cause   Inadequate levels of copper

Signs and symptoms   Terminal leaves become chlorotic and sometimes even dwarfed and deformed and have serrated edges. Older leaves become lighter green than normal, and in severe cases, terminals and lower breaks abort. The cultivar Fransher is especially susceptible to copper deficiency.

Control   Apply copper sulfate to soil surfaces at a rate equivalent to 1.5 lb of $CuSO_4$/1,000 ft^2, or apply copper sprays to foliage. Always include copper in the potting medium (e.g., 1.5 lb of Micromax or 3 lb of Perk/yd^3), or use a periodic micronutrient application of copper. Soil temperatures of 65°F (18°C) or below will contribute to copper deficiency because roots are less able to remove copper from cold soils. Thus, soil temperature should be raised or foliar copper applied during such periods.

Selected Reference

Poole, R. T., and C. A. Conover. 1979. Identification and correction of copper deficiency of *Aglaonema commutatum* 'Fransher'. HortScience 14:187-188.

Copper toxicity

Figure 13

Cause   Foliar applications of copper-containing bactericides

Signs and symptoms   Small, water-soaked areas form at leaf tips or in the parts of the leaves where water or spray accumulates. These spots resemble those caused by bacterial infections when new but generally turn tan. They are very irregular in shape.

Control   When using copper-containing bactericides or nutritional sprays, be sure to test new cultivars for sensitivity. Always use labeled rates and intervals of these products, and never mix them with a product that can lower the pH (such as fosetyl aluminum or vinegar). Water with a pH below 6 can result in copper toxicity.

Selected Reference

Chase, A. R. 1989. Aliette 80WP and bacterial disease control. III. Phytotoxicity. University of Florida, Central Florida Research and Education Center-Apopka, CFREC-A Research Report, RH-89-9.

Dasheen mosaic

Figure 14

Cause   Dasheen mosaic virus (DMV)

Signs and symptoms   DMV is rarely seen on aglaonemas, but symptoms can include mosaic, leaf distortion, and stunting.

Control   DMV can be spread by aphids but is more commonly spread by using infected stock materials during propagation. It is very important to use pathogen-free stock, since the symptoms of DMV are not always noticeable. No chemicals have any known effect on this viral disease.

Selected Reference

Zettler, F. W., M. J. Foxe, R. D. Hartman, J. R. Edwardson, and R. G. Christie. 1970. Filamentous viruses infecting Dasheen and other Araceous plants. Phytopathology 60:983-987.

Erwinia blight

Figures 15, 16, and 17

Cause   *Erwinia carotovora* subsp. *carotovora* or *E. chrysanthemi*

Signs and symptoms   Bacterial blight is typified by watery leaf spots with centers that fall out. Bacterial stem rots caused by *Erwinia* spp. are generally first noticed when cuttings are propagated. The cut end of the stem becomes mushy and foul smelling, and the rooting process stops. The leaves on infected cuttings usually yellow quickly.

Control   Control of bacterial leaf spots or blights can be best accomplished through use of clean propagation material and a watering system that either does not wet the foliage or allows it to dry rapidly. Both antibiotic and copper compounds provide little control of bacterial diseases and are not recommended. Bacterial stem rot is usually not possible to control once started. Use of clean cuttings is the only successful method of cultural control, although some growers have reported dips in streptomycin sulfate as a moderately successful control method.

Selected Reference

McFadden, L. A. 1969. *Aglaonema pictum*, a new host of *Erwinia chrysanthemi*. Plant Dis. Rep. 53:253-254.

Fluoride toxicity

Figure 18

Cause   Excessive amounts of fluoride

Signs and symptoms   Marginal and tip chlorosis is followed by necrosis. Spots are roughly elliptical and can be water soaked, greatly resembling those caused by plant pathogens such as *Erwinia* and *Xanthomonas* spp. Symptoms are also easily confused with those caused by boron toxicity.

Control Fluoride may be present in fertilizer (such as triple superphosphate), irrigation water, and potting medium components (such as perlite and some peat mosses). Some *Aglaonema* cultivars such as Maria are especially sensitive to both boron and fluoride toxicity. If this type of spot appears in an interiorscape, it is more likely caused by nutrient toxicity than by a bacterial pathogen.

Selected Reference

Poole, R. T., and C. A. Conover. 1985. Boron and fluoride toxicity of foliage plants. Univ. of Florida, Agricultural Research and Education Center-Apopka, AREC-A Research Report, RH-85-19.

Fusarium stem rot

Figure 19

Cause *Fusarium subglutinans* (= *F. moniliforme*) and other species of *Fusarium*

Signs and symptoms Fusarium stem rot typically appears as a soft, mushy rot at the base of a cutting or rooted plant. The rotten area frequently has a purplish to reddish margin. *Fusarium* sometimes forms tiny, bright red, globular structures (fruiting bodies) at the stem base of severely infected plants.

Control Thiophanate methyl compounds provide good control of Fusarium stem rot. If stem rot or cutting rot is a problem, treatment of the cuttings with a dip or a post-sticking drench should diminish losses. Remove infected plants from stock areas as soon as they are detected. Since Fusarium stem rot is similar in appearance to Erwinia blight, accurate disease diagnosis is very important prior to applications of pesticides.

Selected Reference

Uchida, J. Y., and M. Aragaki. 1994. Fusarium collar rot and foliar blight of *Aglaonema* in Hawaii. Plant Dis. 78:1109-1111.

Myrothecium leaf spot

Figure 20

Cause *Myrothecium roridum*

Signs and symptoms Myrothecium leaf spot is one of the easiest leaf diseases to diagnose. Leaf spots are frequently found at wounds, although it is common to find no obvious wound and very large (up to 2.5 cm) leaf spots. The spots are usually tan to brown and may have a bright yellow border. Examination of the lower leaf surface shows the black and white fruiting bodies of the pathogen in concentric rings near the outer edge of the spot.

Control Control can be achieved if plant foliage is kept dry and wounding is eliminated. In the absence of this control, fungicides are effective. Both chlorothalonil and mancozeb provide good control of Myrothecium leaf spot on many foliage plants.

Selected References

Chase, A. R. 1983. Influence of host plant and isolate source on Myrothecium leaf spot of foliage plants. Plant Dis. 67:668-671.
Chase, A. R. 1985. 1984 fungicide trials for leaf spot diseases of some foliage plants. Nurserymen's Digest 19(10):70-72.

Pythium root rot

Figure 21

Cause *Pythium splendens* and other *Pythium* spp.

Signs and symptoms Root rot is typified by wilting of plants and yellowing of lower leaves. The roots themselves are brown to black, reduced in mass, and mushy. The outer portion of infected roots can easily be pulled away from the inner core.

Control Using pathogen-free potting medium and pots and growing plants on raised benches can eliminate much of this problem. If fungicides are needed, drenches with etridiazol, etridiazole, fosetyl aluminum, or metalaxyl can aid in control of both Pythium and Phytophthora root rot. Since many times other pathogens are also involved, accurate diagnosis of the cause must be made prior to choosing a fungicide.

Selected References

Chase, A. R., D. D. Brunk, and B. L. Tepper. 1985. Fosetyl aluminum fungicide for controlling Pythium root rot of foliage plants. Proc. Fla. State Hortic. Soc. 98:119-122.
Miller, H. N. 1958. Control of Pythium root rot of Chinese evergreen by soil fumigation. Proc. Fla. State Hortic. Soc. 71:416-419.
Tisdale, W. B., and G. D. Ruehle. 1949. Pythium root rot of aroids and Easter lilies. Phytopathology 39:167-170.

Stunting

Figure 22

Cause Lack of fertilizer

Signs and symptoms Plants develop small, yellow or light green leaves. Roots may develop extensively when soil fertility is low, unlike their sparse condition when a root disease is causing stunting.

Control Check soluble salts of potting medium to ensure that soil fertility is low before applying recommended rates of a complete fertilizer.

Selected Reference

Henny, R. J., A. R. Chase, and L. S. Osborne. 1991. *Aglaonema*. University of Florida, Central Florida Research and Education Center-Apopka, CFREC-A Foliage Plant Research Note, RH-91-2.

Xanthomonas leaf spot

Figure 23

Cause   *Xanthomonas campestris* pv. *dieffenbachiae*

Signs and symptoms   Reddish brown areas with bright yellow margins formed on leaf edges are the most common symptoms. Under wet and warm conditions, bacteria also spread into the leaf centers and lesions expand until they reach a leaf vein.

Control   Minimize water on leaves, and use clean cuttings. Sprays of copper or antibiotic compounds on a weekly basis provide moderate control under some growing conditions. Be careful to use appropriate rates of copper compounds, with water at pH 6.0 or 6.5, since copper toxicity symptoms look similar to those caused by the bacteria.

Selected References

McFadden, L. A. 1962. Two bacterial pathogens affecting leaves of *Aglaonema robelinii.* (Abstr.) Phytopathology 52:20.

McFadden, L. A. 1963. Leaf spot of *Aglaonema robelinii* (Schismaglottis). Ornamental Hort. Rep. 2(1):2-3.

Anthurium

Anthuriums are primarily epiphytes and are native to the tropical Americas. Anthuriums are grown worldwide as a cut-flower crop but are rapidly increasing in status as a potted plant. These plants are generally smaller than those produced for flowers and may have variegated foliage or small, colorful flowers. Large-leafed anthuriums are used either as large specimen plants or in mass plantings for interiorscapes. Anthuriums should be produced under 1,500 to 2,500 ft-c, although this may differ for some of the potted hybrids. Indoors they require at least 150 ft-c to survive, and flowering will probably be rare under these conditions. Greenhouse temperatures should not drop below 65°F (18°C). The most serious disease of anthuriums is caused by a bacterium (see Xanthomonas) that has been reported worldwide. Other diseases are caused by a variety of viruses, nematodes, bacteria, and fungi. Pests of anthurium include mites, thrips, scales, and mealybugs.

Anthracnose

Figure 24

Cause   *Colletotrichum gloeosporioides*

Signs and symptoms   Anthracnose symptoms start as tiny, brownish spots on the flower spadix. Under high humidity, these spots enlarge, appear water soaked, and turn necrotic. Sometimes the entire spadix will turn black as lesions coalesce. The shape of most lesions is, however, angular because of the shape of the spadix tissue. As the disease becomes more severe, masses of orange spores form on necrotic areas. Leaves and spathes are rarely if ever infected.

Control   Research has shown that mancozeb aids in control of this disease. In Hawaii, anthurium breeding programs for cut-flower production routinely select for resistance to this disease. Resistance levels of the potted hybrid anthuriums have not been determined.

Selected References

Aragaki, M., and M. Ishii. 1960. A spadix rot of *Anthurium* in Hawaii. Plant Dis. Rep. 44:865-867.

Aragaki, M., H. Kamemoto, and K. M. Maeda. 1968. Anthracnose resistance in *Anthurium*. Tech. Progress Rep. 169, HITAHR.

Kamemoto, H., M. Aragaki, J. Kunisaki, and T. Higaki. 1977. Breeding for resistance to anthracnose in anthuriums (*Anthurium andraeanum* Lind.). Proc. Trop. Reg. Am. Soc. Hortic. Sci. 19:269-274.

Chimera

Figure 25

Cause   Genetic variability in the plant

Signs and symptoms   Leaves develop yellow variegation that can be mistaken for a viral infection. Plants with these symptoms are usually rare in a group of plants, unlike those infected with a virus, which generally occur in clusters or throughout the planting, depending upon the method of virus spread. Leaf distortion can also occur, especially on some cultivars produced in tissue culture where the plant genetics may become abnormal because of the unusual growing conditions.

Control   If the number of off-type plants is high, a new source of plants should be found. Discard those that are found, but do not miss the opportunity to develop a new selection of the plant. This is one of the oldest ways for new plants to come into the commercial trade.

Copper toxicity

Figure 26

Cause   Copper from foliar applications of bactericides or nutrients

Signs and symptoms   New leaves develop pocking and small, yellow spots. Distortion and burning at the edges can occur.

Control   Not all *Anthurium* cultivars are sensitive to copper, but many of the *A. andraeanum* cultivars used for cut flowers have proved too sensitive for applications of copper-containing bactericides. Test each new cultivar in a small plot before starting broad-scale use of copper-containing products.

Selected Reference

Nishijima, W. T., and D. K. Fujiyama. 1985. Bacterial blight of *Anthurium*. Hawaii Coop. Ext. Serv. HITAHR. Commodity Fact Sheet AN-4(A). 3 pp.

Impatiens necrotic spot

Figure 27

Cause   Impatiens necrotic spot virus (INSV)

Signs and symptoms   Yellow or light green ring spots or a mosaic pattern form on leaves. These could be confused with other diseases (especially Xanthomonas blight) when the symptoms first appear. As the spots age, however, the affected tissue turns brown and dies and shows a more characteristic donut-shaped spot typical of INSV on other hosts.

Control   Thrips are one of the most common means of spread of INSV. Anthuriums are especially prone to thrips feeding injury (leaves and flowers show damage), and a strict control program should be followed. Never propagate from plants with these symptoms, since the new plants have a high likelihood of viral infection. Scout plants thoroughly on a routine basis, and destroy any plants with INSV.

Phytophthora leaf spot, flower blight, and root rot

Figures 28 and 29

Cause   *Phytophthora parasitica*

Signs and symptoms   Phytophthora leaf spot and flower blight are characterized by small, water-soaked spots on leaves and/or spathe tissues. Spots turn black and remain wet appearing as they enlarge. They can encompass the entire flower or leaf under conditions of high temperature and moisture, which are favorable to pathogen development. When conditions become drier or cooler, spots dry and can appear papery but usually remain quite dark in color. Phytophthora root rot shows the same symptoms as many other root rot diseases. Leaves wilt, may turn yellow or pale green, and eventually die. Plants are frequently stunted, and examination of roots reveals their rotted condition. Initial infections of the roots appear as small water-soaked gray or brown areas. These spots can rapidly expand to affect the entire root system. Severely infected plants may have no living roots remaining by the time they are examined.

Control   Prevention is always the best control of a soil-borne pathogen like *Phytophthora*. Use clean pots and potting media, and grow plants on raised benches. Since anthuriums are rarely tolerant of heavy or poorly drained potting media, the appropriate mix is critical. Even regular fungicide applications to infected plants in a heavy potting medium will not control this disease on some potted anthuriums. Fungicides that should aid in control of this root rot include metalaxyl, etridiazole, and fosetyl aluminum. Mancozeb or fosetyl aluminum as a foliar spray on anthuriums may give some control of the leaf and flower blight phases of this disease.

Selected Reference

Aragaki, M. 1986. Floral diseases of *Anthurium*. Hawaii Coop. Ext. Serv. Hort. Dig. 79:6-7.

Pseudomonas leaf spot

Figure 30

Cause   *Pseudomonas cichorii*

Signs and symptoms   Spots start as small, water-soaked areas, which can rapidly enlarge to an inch or more in diameter. The spots may dry out if overhead irrigation is discontinued or the weather turns dry. Sometimes each spot is surrounded by a bright yellow halo.

Control   Minimize foliar wetting from irrigation or rainfall. Other hosts of *P. cichorii* include the majority of other aroids, and care should be taken not to spread disease from one crop to another. Bactericides such as copper compounds or antibiotics are difficult to use on anthuriums effectively, since some cultivars are sensitive to both types of bactericides. Always use pathogen-free plants, and be sure to get an accurate diagnosis, because many leaf diseases look the same when they first appear.

Pythium root rot

Figure 31

Cause   *Pythium splendens* and other *Pythium* spp.

Signs and symptoms   Pythium root rot shows the same symptoms as many other root rot diseases. Leaves wilt, may turn yellow or pale green, and eventually die. Plants are frequently stunted, and examination of roots reveals their rotted condition. Initial infections of the roots appear as small, water-soaked, gray or brown areas. Severely infected plants may have no living roots remaining.

Control   Prevention is always the best control of a soil-borne pathogen like *Pythium*. Use clean pots and potting media, and grow plants on raised benches. Since anthuriums are rarely tolerant of heavy or poorly drained potting media, the appropriate mix is critical. Even regular fungicide applications to infected plants in a heavy potting medium will not control this disease on some potted anthuriums. Fungicides registered for anthuriums that should aid in control of this root rot are the same as those listed for Phytophthora root rot.

Xanthomonas blight

Figures 32 and 33

Cause   *Xanthomonas campestris* pv. *dieffenbachiae*

Signs and symptoms   Symptoms usually start on the leaf margins where the bacterium enters through hydathodes. Spots are first translucent, yellowish, and water soaked. They may take a long time to enlarge, but eventually they can encompass the entire leaf margin, invade the center of the leaf, and even cause leaf drop. Mature spots are black and usually surrounded by a bright yellow halo. If the anthurium becomes systemically infected, the plant will show signs of yellowing, stunting, and loss of lower leaves and will eventually die. Natural infection of the spadix (resembling anthracnose) has been reported in Hawaii. Systemic infections may also cause leaf spots, which form anywhere on the leaves.

Control Use of bactericides for control of even the foliar phase of Xanthomonas blight is rarely effective. *Anthurium andraeanum* cultivars typically show phytotoxic responses to both streptomycin sulfate and copper compounds, although these products can be used safely on some of the potted anthurium hybrids. Pathogen resistance to copper has been reported widely. Avoidance of this disease is the most effective control. Scout the crop routinely and frequently to detect early symptoms of Xanthomonas blight. Some growers report effective control by removing symptomatic leaves, although this method has obvious drawbacks for potted foliage producers. Limit overhead irrigation to reduce pathogen spread, and keep in mind that most of the commonly produced aroids (*Dieffenbachia*, *Aglaonema*, and *Syngonium*) are also hosts of this pathogen. Disinfest tools with 70% alcohol for at least 3 minutes, or use another suitable disinfestant (such as quaternary ammonium). Disease management by fertilizer is not acceptable, since rates that reduce bacterial leaf spot also stop flower production.

Selected References

Alvarez, A. 1988. Proceedings of the First Anthurium Blight Conference. HITAHR, University of Hawaii at Manoa. 46 pp.

Alvarez, A. 1990. Proceedings of the Third Anthurium Blight Conference. HITAHR, University of Hawaii at Manoa. 40 pp.

Chase, A. R. 1989. Effect of fertilizer rate on susceptibility of *Anthurium andraeanum* to *Xanthomonas campestris* pv. *dieffenbachiae*, 1988. Biological and Cultural Tests 4:80.

Chase, A. R. 1990. Effect of nitrogen source on growth and susceptibility of *Anthurium* hybrids to *Xanthomonas campestris* pv. *dieffenbachiae*. University of Florida, Central Florida Research and Education Center-Apopka, CFREC-A Research Report, RH-90-20.

Chase, A. R. 1992. Effect of potassium rate on susceptibility of *Anthurium andraeanum* to *Xanthomonas campestris* pv. *dieffenbachiae*, 1990. Biological and Cultural Tests 7:115.

Fernandez, J. A., and W. T. Nishijima. 1989. Proceedings of the Second Anthurium Blight Conference. HITAHR, University of Hawaii at Manoa. 55 pp.

Hayward, A. C. 1972. A bacterial blight of anthurium in Hawaii. Plant Dis. Rep. 56:904-908.

Nishijima, W. T., and D. K. Fujiyama. 1985. Bacterial blight of *Anthurium*. Hawaii Coop. Ext. Serv. HITAHR. Commodity Fact Sheet AN-4(A). 3 pp.

Aphelandra (Zebra plant)

Zebra plants and their relatives are herbaceous to semiwoody plants native to tropical America. Most are grown for sale as small container plants because of their colorful foliage and bracts (the flowers are relatively inconspicuous). Plants are produced under approximately 1,500 to 2,000 ft-c and will tolerate a minimum of 150 ft-c indoors. Night temperature during production should be no lower than 65°F (18°C). Flowering occurs best at 1,000 ft-c during the long days of summer but requires 2,500 ft-c during shorter winter days. These plants are hosts of a wide variety of fungal pathogens that cause leaf, stem, and root diseases as well as several idiopathic disorders. Several pests are common on zebra plants, including mealybugs, aphids, and tarsonemid mites. Snails and slugs also damage these plants.

Bendiocarb phytotoxicity

Figure 34

Cause Application of bendiocarb to the potting medium

Signs and symptoms A single application of bendiocarb to the potting medium results in necrotic spots, tip necrosis, leaf cupping, and distortion.

Control Always follow pesticide labels for applications rates, intervals, and especially sites. Products meant to be applied to leaves are not always safe when applied to the roots via the potting medium.

Selected Reference

Osborne, L. S., and A. R. Chase. 1986. Phytotoxicity evaluations of Dycarb on selected foliage plants. University of Florida, Agricultural Research and Education Center-Apopka, AREC-A Research Report, RH-86-12.

Chlorosis

Figure 35

Cause Low fertilizer levels

Signs and symptoms Plants develop small leaves that are light green or yellow. Eventually, they cease to grow. Root systems are generally well developed and larger than would be expected given the degree of top growth.

Control Apply fertilizer regularly at recommended rates. Be sure to test the soluble salts of the potting medium before applying fertilizer to plants that are yellow and/or stunted. There are a number of soilborne fungi and nematodes that can cause these symptoms. In addition, on some plants symptoms of overfertilization can appear similar to those of underfertilization.

Selected Reference

Poole, R. T., A. R. Chase, and L. S. Osborne. 1991. *Aphelandra*. University of Florida, Central Florida Research and Education Center-Apopka, CFREC-A Foliage Plant Research Note, RH-91-4.

Corynespora leaf spot

Figure 36

Cause *Corynespora cassiicola*

Signs and symptoms Leaf spots start on leaf edges, tips, and sometimes centers, particularly near the pot edges and potting medium and at wound sites. They are dark brown to black and often appear wet. This disease can be a serious problem on cuttings rooted under mist and on bottom leaves of potted plants. Lesions expand rapidly and may reach 5 cm or more when conditions are favorable. There is rarely any halo surrounding lesions on zebra plants. The white Apollo zebra plant cultivar is more susceptible than the dark green Dania cultivar. Corynespora leaf spot of zebra plant is usually a problem only during the propagation phase, since only then are plants kept under very high moisture and humidity conditions. Apparently healthy cuttings may develop severe symptoms of disease.

Control Elimination of overhead water can control this disease. In situations in which chemicals are needed, both mancozeb and chlorothalonil provide excellent control.

Selected References

Chase, A. R. 1981. Comparison of *Myrothecium* sp. and *Corynespora cassiicola* leafspots on two cultivars of *Aphelandra squarrosa* Nees. Proc. Fla. State Hortic. Soc. 94:115-116.

McRitchie, J. J., and J. W. Miller. 1973. Corynespora leaf spot of Zebra plant. Proc. Fla. State Hortic. Soc. 86:389-390.

Crinkle leaf

Figure 37

Cause Unknown

Signs and symptoms Aphelandra leaf crinkle is typified by pronounced stunting of terminal growth, reduced leaf size, and downward puckering and reflexing of leaves. Symptoms are most severe under high light levels (1,500 ft-c and higher) but cannot be reproduced with light alone. Affected plants continue to produce symptomatic leaves regardless of cultural conditions. In addition, production of rooted cuttings from crinkle-affected plants is reduced in both quality of cuttings available and ability of the cutting to root.

Control The disorder is most severe under high light levels and high temperature. No bacteria, fungi, or viruses have been found associated with this disorder, and pesticides have no effect. Plants do not respond to micronutrients or proper environmental conditions. Eliminate stock plants with this problem, and maintain proper light levels and temperatures.

Selected References

Henley, R. W. 1982. Aphelandra leaf crinkle-Aphelandra stunt—A report from Florida. University of Florida, Agricultural Research Center-Apopka, ARC-A Research Report, RH-82-1.

Kamp, M., A. E. Nightingale, and R. W. Toler. 1981. Leaf crinkle disease of *Aphelandra* in Texas. Plant Dis. 65:687-688.

Kutilakesa canker

Figure 38

Cause *Kutilakesa pironii*

Signs and symptoms Galls are oriented longitudinally as well as at nodes and the cut end of stems. Galls formed on leaf midveins and petioles are about 6 mm in diameter. Perithecia of *Nectriella pironii* (sexual stage) have been found associated with the galls in addition to the sporodochia of *K. pironii* (asexual stage). Wounded tissue appears to be a requirement for infection by *K. pironii*. Symptoms of the disease appear within 3 to 6 weeks after infection and continue to develop for up to 9 months.

Control Predacious mycophagous mites (*Tyrophagus* and *Peloribates* spp.) are frequently associated with galled tissue and are perhaps responsible for the spread of conidia to new sites. Contamination of cutting instruments used during plant propagation also leads to disease spread. Treatment of cutting tools with 70% alcohol or a quaternary ammonium compound will reduce spread of conidia throughout a block. Chemical controls have not been investigated for this disease.

Selected References

Alfieri, S. A., Jr., J. F. Knauss, and C. Wehlburg. 1979. A stem-gall and canker-inciting fungus new to the United States. Plant Dis. Rep. 63:1016-1020.

Alfieri, S. A., Jr., C. L. Schoulties, and N. E. El-Gholl. 1980. *Nectriella* (*Kutilakesa*) *pironii*, a pathogen of ornamental plants. Proc. Fla. State Hortic. Soc. 93:218-219.

Myrothecium leaf spot

Figures 39 and 40

Cause *Myrothecium roridum*

Signs and symptoms Leaf spots caused by this pathogen are similar in appearance to those caused by *C. cassiicola* when viewed from the upper leaf surface. Leaf undersides generally reveal the presence of the fungal fruiting bodies, which are formed in concentric rings within the dead spots. These fruiting bodies are irregularly shaped and black, have a white fringe, and are about the size of a pin head.

Control The same chemical controls used for Corynespora leaf spot apply to Myrothecium leaf spot. Myrothecium leaf spot is most severe when temperatures are between 70 and 85°F (21 and 30°C) but can be a problem throughout the year in greenhouses. Temperatures above 90°F (32°C) greatly inhibit *Myrothecium* and make chemical application less important. The host range of *M. roridum* is very wide and includes aglaonema, dieffenbachia, ferns, ficus, and spathiphyllum. Be sure to scout all susceptible hosts for symptoms of Myrothecium leaf spot when designing an effective control program. Fertilizer manipulation does not affect severity of Myrothecium leaf spot on zebra plant.

Selected References

Chase, A. R. 1981. Comparison of *Myrothecium* sp. and *Corynespora cassiicola* leafspots on two cultivars of *Aphelandra squarrosa* Nees. Proc. Fla. State Hortic. Soc. 94:115-116.

Chase, A. R. 1985. 1984 Fungicide trials for leaf spot diseases of some foliage plants. University of Florida, Agricultural Research and Education Center-Apopka, AREC-A Research Report, RH-85-16.

Chase, A. R., and R. T. Poole. 1986. Effect of nutrition on growth of *Aphelandra squarrosa* and severity of Myrothecium leaf spot. University of Florida, Agricultural Research and Education Center-Apopka, AREC-A Research Report, RH-86-3.

Phytophthora stem rot

Figure 41

Cause   *Phytophthora parasitica*

Signs and symptoms   Stem rot usually starts at the soil line and causes a blistering of the stem surface. These areas are black and slightly mushy and can extend from the base of the stem up into the petioles of lower leaves. Complete collapse of the plant is common.

Control   Control should be based on use of pathogen-free cuttings, pots, and potting media, since the pathogen is easily introduced in any of these ways. Disease is most severe at temperatures between 81 and 91°F (about 27 and 33°C). Chemicals that provide control of this disease include metalaxyl and fosetyl aluminum.

Selected References

Chase, A. R. 1981. Phytophthora stem rot of *Aphelandra squarrosa*. Plant Dis. 65:921-922.

Chase, A. R. 1988. Efficacy of Aliette 80WP in controlling Pythium root rot of *Begonia* and Phytophthora stem rot of *Aphelandra*. University of Florida, Central Florida Research and Education Center-Apopka, CFREC-A Research Report, RH-88-5.

Chase, A. R. 1990. Effect of Osmocote rate on severity of Phytophthora stem rot of *Aphelandra squarrosa* 'Dania'. University of Florida, Central Florida Research and Education Center-Apopka, CFREC-A Research Report, RH-90-23.

Chase, A. R., and T. A. Mellich. 1992. Fungicide tests for control of *Phytophthora* and *Pythium* diseases on ornamentals. University of Florida, Central Florida Research and Education Center-Apopka, CFREC-A Research Report, RH-92-9.

Tomato spotted wilt

Figure 42

Cause   Tomato spotted wilt virus (TSWV)

Signs and symptoms   Light brown necrosis of the major veins appears on recently mature terminal leaves. Extensive blackening of the veins can develop, with epinasty, distortion, and premature abscission.

Control   Thrips are one of the most common ways for spreading both INSV (Impatiens necrotic spot virus) and TSWV. A strict thrips control program should be followed if TSWV is found on zebra plants or other hosts of the virus. Never propagate from plants with these symptoms, since the new plants have a high likelihood of viral infection. Scout plants thoroughly on a routine basis, and destroy any plants with TSWV.

Selected Reference

Halliwell, R. S., and L. W. Barnes. 1987. Tomato spotted wilt virus infection of commercial *Aphelandra* sp. J. Environ. Hort. 5:120-121.

Araucaria (Norfolk Island pine)

Norfolk Island pine is the only coniferous plant commonly produced for interior foliage plant use. Interior uses include dish gardens, desk top plants, floor plants, and large specimen trees. Norfolk Island pine are produced under full sun and are sometimes acclimatized under 5,000 to 6,000 ft-c, with a minimum temperature of 45 to 50°F (7 to 10°C). Interior light requirements are 100 to 150 ft-c or greater. When plants are small, they may be subject to fungal root and stem diseases. However, once past the seedling stage, they are relatively disease resistant with the exception of stem cankers caused by fungi such as *Cylindrocladium* and *Phytophthora* spp. Common pests include mealybugs and scales.

Anthracnose

Figure 43

Cause   *Colletotrichum derridis*

Signs and symptoms   Needle necrosis or anthracnose begins as small necrotic areas on the needles. Large portions of branches may turn brown, and needles abscise readily. Fruiting bodies (acervuli) of the pathogen form in the necrotic areas and appear as tiny, black specks, which are easily seen with the naked eye.

Control   Elimination of most exposure to overhead irrigation or rainfall can control this disease. Preventive applications of thiophanate methyl or chlorothalonil can reduce disease severity on this plant.

Selected Reference

Ridings, W. H. 1973. Colletotrichum needle necrosis of Norfolk Island pine. Proc. Fla. State Hortic. Soc. 86:418-421.

Arecaceae (Palms)

Most palms are native to the tropical and subtropical regions of the world. They occur in the interiorscape in various situations ranging from dish gardens and terrariums to large specimen trees. Palm cultivation depends

**TABLE 6. Suggested light and nutritional levels
for production of some indoor palms[a]**

Palm species	Light intensity (ft-c)	Fertilizer requirement (lbs/1,000 ft²/yr)		
		N	P₂O₅	K₂O
		N	P_2O_5	K_2O
Caryota spp.	3,000–5,000	34	11	23
Chamaedorea cataractarum	4,000–6,000	34	11	23
Chamaedorea elegans	1,500–3,000	28	9	19
Chamaedorea Ernesti-Augusti	1,500–3,000	28	9	19
Chamaedorea erumpens	3,000–6,000	34	11	23
Chamaedorea microspadix	3,000–5,000	28	9	19
Chamaedorea Seifrizii	4,000–6,000	34	11	23
Chrysalidocarpus lutescens	4,000–6,000	34	11	23
Howea Belmoreana	3,000–5,000	34	11	23
Howea Forsterana	3,000–5,000	34	11	23
Livistona chinensis	4,000–6,000	34	11	23
Phoenix Roebelenii	4,000–6,000	34	11	23
Rhapis excelsa	2,000–5,000	34	11	23
Rhapis humilis	2,000–5,000	34	11	23
Rhapis 'Thai-Dwarf'	2,000–5,000	34	11	23

[a] Adapted from Burch, D., R. Atilano, and J. Reinert. 1983. Indoor palm production guide for commercial growers. Univ. of Florida, Ornamental Hort. Fact Sheet, OHC-8.

upon species with production light levels ranging from full sun to 1,500 to 3,000 ft-c (parlor palm) and 4,000 to 6,000 ft-c (areca palm). See Table 6 for more specific recommendations for a variety of indoor palms. *Chamaedorea* palms tolerate indoor light levels of 75 ft-c, while most others require at least 150 ft-c. Palms grown as foliage plants are subject to fungal root rots, especially when young, and to fungal leaf spots at any time. Mites, mealybugs, scales, and thrips are common pests.

Anthracnose

Figure 44

Cause   *Colletotrichum gloeosporioides*

Signs and symptoms   Spots can be small, water-soaked speckles or large, necrotic and chlorotic lesions that are circular to irregular in shape. They are usually tan to black with a bright yellow halo and often coalesce as symptoms progress. On some palms, lesions are angular because they are confined to the area between leaf veins. On other hosts, a frogeye spot may develop.

Control   Any method that keeps foliage dry reduces the potential for infection and spread of the pathogen. Do not use overhead irrigation or expose plants to rainfall if possible. Improve rapid drying of the foliage by spacing plants adequately, and rogue severely infested flats. Preventive applications of many fungicides can greatly reduce incidence of the disease.

Boron deficiency

Figure 45

Cause   Insufficient boron (B) in the soil

Signs and symptoms   Early boron deficiency symptoms include chlorotic new leaves that are usually malformed

and fail to expand normally. Leaf margins are often necrotic, and in more severe cases, entire leaflets may be necrotic. In the most severe cases, only necrotic petiole stubs will emerge and death of the meristem will follow.

Control   Boron is readily leached from most soils. Soil or foliar applications of sodium borate or boric acid will treat boron-deficient palms. Boron should not be applied indiscriminately since toxicity can occur at fairly low levels.

Calonectria or Cylindrocladium leaf spot

Figure 46

Cause   *Calonectria theae, C. colhounii, C. crotalariae*; and *Cylindrocladium floridanum, C. pteridis* (= *C. macrosporum*)

Signs and symptoms   Leaf spots are characteristically grayish brown, dark brown, or nearly black, circular to irregular, and sometimes have a dark border. Young spots are brown, circular, frequently surrounded by a yellow band, and become dark brown to black with grayish brown centers. Advanced stages of the disease are characterized by coalescing leaf spots and chlorosis and necrosis of leaflet margins and tips.

Control   Any cultural practice that reduces free moisture on leaves (increased spacing among plants, covered greenhouses, and drip irrigation) will reduce disease incidence and spread. Prompt removal and destruction of infected leaves, especially blighted leaves, is crucial. Check fungicide labels for this disease and the specific palm concerned.

Selected References

Leahy, R. M. 1989. Cylindrocladium leaf spot on palms. Fla. Dept. Agric. & Cons. Serv. Div. of Plant Industry. Pl. Path. Circ. No. 315.

Uchida, J. Y., and M. Aragaki. 1992. Calonectria leaf spot of *Howea forsterana* in Hawaii. Plant Dis. 76:853-856.

Copper toxicity

Figure 47

Cause   Foliar applications of copper products

Signs and symptoms   Brown, elliptical spots form on the pinnae (leaflets) and sometimes resemble fungal leaf spot disease. Fronds may have numerous necrotic spots along their entire length.

Control   Spots sometimes are caused by foliar applications of micronutrients or copper-containing fungicides. Both copper chelate and cupric hydroxide fungicides can cause phytotoxicity on small areca palms. The concentration of copper used in blended micronutrient fertilizers is not considered a potential problem if they are used at recommended rates.

Selected Reference

Chase, A. R., and R. T. Poole. 1984. Influence of foliar applications of micronutrients and fungicides on foliar necrosis and leaf spot disease of *Chrysalidocarpus lutescens*. Plant Dis. 68:195-197.

Damping-off

Figure 48

Cause *Fusarium, Rhizoctonia, Pythium,* and *Phytophthora* spp.

Signs and symptoms Poor germination, blackening of roots, or mushiness can be followed by yellowing, wilting, and loss of the plant or seedling. An overall poor stand results, which makes repotting or discarding necessary.

Control These organisms usually have wide host ranges and frequently affect many plant species in the same nursery. Always use clean palm seed, new potting media, and clean pots, and whenever possible, grow plants on raised benches to limit exposure to native pathogens in the soil under the pots. The soil moisture should be maintained as low as possible to reduce pathogen growth without reducing plant growth. Preventive drenches with a variety of fungicides are often chosen by growers as extra insurance against damping-off of expensive seeds. Since different fungicides must be used to control *Fusarium* and *Rhizoctonia* than those used for *Pythium* and *Phytophthora*, accurate diagnosis of the specific pathogens causing the root rot is critical.

Selected Reference

Schulman, J. F. 1971. Etiology of a disease complex in *Chamaedorea elegans*. Univ. of Florida, Thesis. 27 pp.

Fertilizer toxicity (high soluble salts)

Figure 49

Cause High soil soluble salts caused by excessive fertilization, use of high-salt fertilizers, or use of saline irrigation water

Signs and symptoms Necrotic leaflet tips on older leaves, yellow new foliage, and stunting are common. Roots will often have necrotic tips or more extensive necrosis. If fertilizer accumulates next to a stem, the plant may collapse at that point.

Control Low-salt fertilizers should be used and only at recommended rates. Some leaching of the soil should occur at each irrigation. If the problem is caused by saline irrigation water and a cleaner water source cannot be found, the soil should not be allowed to dry out and/or only salt-tolerant species of palms should be grown. If the condition already exists, the soil should be leached thoroughly several times to remove excess salts. Never pile fertilizer near the stem of any plant.

Fluoride toxicity

Figure 50

Cause Excess fluoride (F)

Signs and symptoms Foliar necrosis frequently starts as small spots and then progresses until entire leaflets are necrotic. Frequently, plants show a dark brown tip necrosis.

Control Avoid water, medium components (perlite), and fertilizers (superphosphate) that contain fluoride. Leach thoroughly to remove fluoride from the soil. Chamaedoreas are moderately sensitive to fluoride. When minimal amounts of fluoride are present in the medium solution, damage to the plant can usually be prevented by maintaining the medium pH between 6.0 and 7.0, which reduces availability of fluoride.

Selected References

Poole, R. T., and C. A. Conover. 1981. Dolomite and fluoride affect foliar necrosis of *Chamaedorea seifrizii* and *Chrysalidocarpus lutescens*. Proc. Fla. State Hortic. Soc. 94:107-109.
Poole, R. T., and C. A. Conover. 1981. Influence of fertilizer, dolomite, and fluoride levels on foliar necrosis of *Chamaedorea elegans* Mart. HortScience 16:203-205.

Helminthosporium leaf spot

Figures 51 and 52

Cause *Bipolaris* spp., *Exserohilum rostratum,* and *Phaeotrichoconis crotalariae*

Signs and symptoms Lesions are usually 2 to 5 mm long, reddish brown to black, and found all over the frond surface. A yellow halo (margin) frequently surrounds a lesion. Under optimum conditions, lesions coalesce and form large, irregularly shaped necrotic areas on leaf tips and margins. Sometimes, the unexpanded newest leaf may become entirely necrotic.

Control Very little disease develops on plants that are not exposed to overhead irrigation or rainfall. Although many growers have reported that poorly fertilized plants are more susceptible to Helminthosporium leaf spot than well-fertilized plants, research has not established this relationship. Plants grown in full sun are more severely affected by this disease than those grown under reduced light (50% shade). Optimum disease control can be achieved with weekly applications of chlorothalonil, mancozeb, or iprodione. Check labels for specific palms.

Selected References

Chase, A. R. 1982. Influence of irrigation method on severity of selected fungal leaf spots of foliage plants. Plant Dis. 66:673-674.
Chase, A. R. 1982. Dematiaceous leaf spots of *Chrysalidocarpus lutescens* and other palms in Florida. Plant Dis. 66:697-699.
Chase, A. R. 1990. 1990 fungicide trials for control of *Alternaria, Helminthosporium, Phytophthora,* and *Rhizoctonia* diseases of ornamentals. University of Florida, Central Florida Research and

Education Center-Apopka, CFREC-A Research Report, RH-90-25.

Chase, A. R., and R. T. Poole. 1983. The feasibility of controlling areca palm leaf spot by altering host nutrition. Proc. Fla. State Hortic. Soc. 96:278-280.

Chase, A. R., and R. T. Poole. 1984. Influence of foliar applications of micronutrients and fungicides on foliar necrosis and leaf spot disease of *Chrysalidocarpus lutescens*. Plant Dis. 68:195-197.

Iron toxicity

Figure 53

Cause   Foliar applications of iron products

Signs and symptoms   Brown, elliptical spots form on the pinnae (leaflets) and are sometimes confused with fungal leaf spots. Fronds may have numerous necrotic spots along their entire length.

Control   Spots sometimes are caused by foliar applications of micronutrients or iron-containing fungicides. Both iron chelate and ferbam (a carbamate fungicide with iron) result in phytotoxicity and should not be used on small areca palms. The concentration of iron used in blended micronutrient fertilizers is not considered a potential problem if used at recommended rates.

Selected Reference

Chase, A. R., and R. T. Poole. 1984. Influence of foliar applications of micronutrients and fungicides on foliar necrosis and leaf spot disease of *Chrysalidocarpus lutescens*. Plant Dis. 68:195-197.

Magnesium deficiency

Figure 54

Cause   Insufficient magnesium (Mg) in the soil

Signs and symptoms   The oldest leaves usually have broad, chlorotic bands along the margins; the chlorosis starts at the leaflet tips and expands toward the rachis as the deficiency progresses. In severe cases, only the rachis and adjacent portions of the leaflets remain green on the oldest leaves, but younger leaves show progressively wider bands of green along the centers of leaves.

Control   Magnesium is readily leached from sandy soil and other soils having little cation exchange capacity. High levels of potassium or calcium in the soil also can induce magnesium deficiency. Magnesium deficiency is difficult to correct once symptoms are present. It is best prevented by amending all container media with dolomite. Foliar magnesium sprays are generally ineffective in treating magnesium deficiency because they supply very small amounts of magnesium relative to the amount required by palms.

Phytophthora blight

Figures 55 and 56

Cause   *Phytophthora nicotianae* (= *P. parasitica*), *P. palmivora*

Signs and symptoms   Diseases caused by *Phytophthora* include seedling blights and damping-off; trunk, crown, and root rots; leaf spots, blights, and petiole rots; and apical tip, bud, or heart rot followed by the death of the plant. Phytophthora root and stem rot occurs mostly during the summer months and is typified by severe loss of roots and wilting of the tops. Roots are blackened and their cortex is easily removed from the central core. The symptoms on the upper portions may be confined to loss of stems starting with lesions near the soil line and yellowing of leaves on these stems or can include discrete lesions on the stems. Lesions are black and sunken and can appear on portions as high above the soil as 0.3 m.

Control   Clean plant or seed sources are vital for disease avoidance. Early removal and destruction of infected plants will reduce inoculum levels and decrease the incidence of new infections. Free water favors the development of *Phytophthora*. Once disease occurs, severely affected palms should be rogued from either the nursery or the landscape and burned. Curative efforts with foliar fungicides (fosetyl aluminum and metalaxyl) have been effective in limiting disease spread experimentally.

Selected References

Atilano, R. A. 1982. Phytophthora bud rot of Washingtonia palm. Plant Dis. 66:517-519.

Nagata, N. M., and M. Aragaki. 1989. Etiology and control of Phytophthora leaf blight of golden-fruited palm. Plant Dis. 73:661-663.

Pink rot (Gliocladium stem blight)

Figure 57

Cause   *Gliocladium vermoeseni*

Signs and symptoms   This disease is an invasive rot that can attack the bud tissues, petioles, leaf blades, and trunks. Spots are often associated with gummy exudates. Older fronds die prematurely, necrotic streaks appear from the rachis base, and pinnae turn chlorotic. The pathogen readily produces dusty masses of orange to pink conidia. In severe infections, many stems are girdled and die, making the potted plants unmarketable. Removal of symptomatic fronds reveals stem infections, which are dark brown and irregularly shaped, sometimes with chlorotic margins.

Control   Disease appears to be most severe during the winter months. Since wounding facilitates infection, only completely dead leaves should be removed from palms with Gliocladium blight. Yellow leaves should be removed only when temperatures are at least 85°F (30°C), preferably 90°F (32°C), to reduce chances of infection. Applications of mancozeb at either 7- or 14-day intervals gave excellent disease control during the summer months in Florida on *Chamaedorea* palms.

Selected References

Atilano, R. A., W. R. Llewellyn, and H. M. Donselman. 1980. Control of *Gliocladium* in Chamaedorea palms. Proc. Fla. State Hortic. Soc. 93:194-195.

Reynolds, J. E. 1964. Gliocladium disease of palm in Dade County, Florida. Plant Dis. Rep. 48:718-720.

Potassium deficiency

Figure 58

Cause   Insufficient potassium (K) in the soil

Signs and symptoms   Leaflets of some palms are mottled with yellowish spots that are translucent when viewed from below. In other palms, symptoms appear on older leaves as a marginal or tip necrosis of the leaflets with little or no yellowish spotting present. Early symptoms in some palms appear as necrotic streaks within the leaflets. In more severely deficient palms, marginal or marginal and tip necrosis will be present. The most severely affected leaves or leaflets will be completely necrotic and withered in appearance.

Control   Regular applications of potassium fertilizers will prevent potassium deficiency and treat palms already deficient in potassium. On sandy soils or on those having little cation exchange capacity, potassium fertilizer should be applied in coated, controlled-release forms to prevent rapid loss caused by leaching. Fertilizers with a 3N-1P-3K-1Mg ratio appear to be best for palms.

Pseudocercospora leaf spot

Figure 59

Cause   *Pseudocercospora rhapisicola* (= *Cercospora rhapisicola*)

Signs and symptoms   Leaf spots begin as very tiny, faintly yellow or light green flecks. These spots expand into elliptical or circular spots along parallel veins. Larger spots surrounded by pinpoint satellite speckles are typical in later stages of disease development. These areas are composed of slightly raised, dark brown to reddish brown flecks and spots surrounded by chlorotic tissue that gradually turns brown.

Control   Close adherence to a program of sanitation and eradication of infected leaves is the key to *Pseudocercospora* disease control. Complete removal of diseased leaves followed by regular inspection and removal of new diseased leaves will reduce inoculum levels. Maintaining sanitation and keeping plants under solid-cover greenhouses should make chemical treatment unnecessary.

Selected References

Tominaga, T. 1965. Brown leaf spot of *Rhapis flabelliformis* L'Her. caused by *Cercospora rhapisicola* sp. nov. Trans. Mycol. Jpn. 5:57-59.

Uchida, J. Y., and N. M. Nagata. 1989. Pseudocercospora leaf spot of Rhapis palm. Univ. of Hawaii, HITAHR Brief No. 076.

Pseudomonas blight

Figure 60

Cause   *Pseudomonas albo-precipitans* (*P. avenae*)

Signs and symptoms   The first symptoms of this disease are small, water-soaked, translucent areas running along leaf veins. Mature lesions are brown to black, have a chlorotic halo, and range from 1 mm wide and up to 5 cm long. In many cases, the initial infection appears to occur at leaf margins.

Control   Optimal temperatures for disease development are between 65 and 90°F (18 and 32°C). No chemical controls have been investigated for this disease, although copper or antibiotic products may be somewhat effective. Eliminating overhead irrigation and removing symptomatic leaves or entire plants are the recommended cultural controls.

Selected Reference

Knauss, J. F., J. W. Miller, and R. J. Virgona. 1978. Bacterial blight of fishtail palm, a new disease. Proc. Fla. State Hortic. Soc. 91:245-247.

Sclerotinia blight

Figure 61

Cause   *Sclerotinia homeocarpa*

Signs and symptoms   Sclerotinia blight of palms occurs on small seedlings up to 0.8 m tall. Foliar blighting is accompanied by gray to white mycelium, which frequently covers overlapping pinnae of affected fronds. Individual lesions are irregular in shape and surrounded by a water-soaked band of tissue. Lesions eventually turn tan to gray with a dark brown border.

Control   Sclerotinia blight of Areca palms is most severe on plants less than 0.8 m tall, since they are densely planted (up to 100 seeds in an 20-cm pot), creating ideal conditions for infection and disease development. Increase plant spacing, water early in the day, and remove all symptomatic plants to control this disease.

Sunburn

Figure 62

Cause   Exposure of shade-grown foliage to high light intensities

Signs and symptoms   Large necrotic areas are visible on the upper surfaces of leaves, usually in the centers of leaves or leaflets rather than on leaf tips or margins. Affected foliage on adjacent leaves will often have the same directional orientation.

Control Sunburn can be prevented by growing palms in full sun if they are to be used in the landscape. Acclimatization of palm leaves involves replacement of the entire canopy by leaves adapted to the new higher light intensity. Individual shade-grown leaves cannot adapt to higher light intensities.

Zinc deficiency

Figure 63

Cause Inadequate levels of zinc (Zn) in the soil

Signs and symptoms New leaves show interveinal chlorosis in most zinc-deficient palms. In more severely deficient palms, leaflet tips become necrotic and the necrosis increases until only necrotic petiole stubs remain. Death of the meristem will occur if corrective treatment is not given.

Control Zinc deficiency can be prevented by using fertilizers containing zinc. Treatment of deficient palms with zinc sulfate applied to the soil or foliage is usually effective. High soil pH can precipitate zinc and render it unavailable to palms.

Selected Reference

Marlatt, R. B., and J. J. McRitchie. 1979. Zinc deficiency symptoms of *Chrysalidocarpus lutescens*. HortScience 14:620-621.

Begonia

Begonias are native to tropical and subtropical regions throughout the world. Those most commonly classed as foliage plants are the Rex begonias. Many more are used as flowering potted plants or bedding plants. They are frequently used in dish gardens or in mass plantings in interiorscapes. Commercial propagation of Rex begonias relies on use of leaf cuttings or tissue-cultured plantlets. Light levels of about 2,000 to 2,500 ft-c and temperatures between 62 and 85°F (16 and 30°C) produce good-quality foliage begonias. Indoors they require 150 to 500 ft-c for best appearance. Diseases caused by fungi are common as well as bacterial blight caused by *Xanthomonas campestris* pv. *begoniae*. Mites, scale insects, and mealybugs can be problems.

Botrytis blight

Figure 64

Cause *Botrytis cinerea*

Signs and symptoms Botrytis leaf blight usually appears on lower leaves of cuttings in contact with the potting medium. The water-soaked area may enlarge rapidly to encompass a large portion of the leaf or even the entire cutting. The infected area dies and turns dark brown to black with age. When night temperatures are cool, day temperatures warm, and moisture high, the fungus readily sporulates on leaves, covering them with grayish brown dusty masses of conidia.

Control Cultural control methods that improve foliage drying and reduce moisture condensation on foliage during the nights reduce the need for fungicide application. Mancozeb, iprodione, chlorothalonil, and vinclozolin are effective for control of *Botrytis* on begonia.

Fusarium stem rot

Figure 65

Cause *Fusarium solani* and *F. oxysporum*

Signs and symptoms Lower leaves wilt and turn yellow. Stem bases are dry and have cankers, sunken areas that are usually brown and may be mushy.

Control Use cuttings from pathogen-free stock or from tissue culture. New or sterilized pots and potting medium are essential. Chemical controls for begonia include thiophanate methyl, ferbam, and chlorothalonil.

Myrothecium leaf spot

Figures 66 and 67

Cause *Myrothecium roridum*

Signs and symptoms Spots generally appear at leaf edges and tips or at broken leaf veins. Dead areas are dark brown and initially appear water soaked. Examination of the lower leaf surface generally reveals sporodochia, which are irregularly shaped, black with a white fringe of mycelium, and formed in concentric rings within the dead areas.

Control There are numerous foliage plant hosts of *Myrothecium,* including lipstick vine, aglaonema, dieffenbachia, ficus, and ferns. Using fungicides when temperatures are between 70 and 85°F (21 and 30°C), minimizing wounding, and fertilizing at recommended levels minimize the severity of Myrothecium leaf spot. Chlorothalonil and mancozeb provide excellent control of this disease.

Selected References

Chase, A. R. 1982. Chemical control of Myrothecium crown rot and leaf spot of Rex begonia. University of Florida, Agricultural Research Center-Apopka, ARC-A Research Report, RH-82-10.

Chase, A. R. 1983. Influence of host plant and isolate source on Myrothecium leaf spot on foliage plants. Plant Dis. 67:668-671.

Powdery mildew

Figure 68

Cause *Oidium begoniae*

Signs and symptoms Spots of frosty white growth appear on leaves. The powdery coating can form circular areas as single spots up to 1 cm in diameter or the spots can join to cover the entire leaf. Many spots are found on leaf undersides. The disease is most common during the drier periods of the year and in interiorscapes.

Control   Disease is most severe at 70°F (21°C) and more or less ceases when temperatures exceed 82°F (28°C). The disease apparently does not cause serious losses on Rex begonias, since many growers do not apply fungicides during an outbreak. Triforine, fenarimol, and triadimefon are effective for powdery mildew control on begonia. Research from New Zealand reported excellent control with dinocap.

Selected Reference

Quinn, J. A., and C. C. Powell, Jr. 1982. Effects of temperature, light and relative humidity on powdery mildew of begonia. Phytopathology 72:480-484.

Pythium root rot

Figure 69

Cause   *Pythium splendens* and sometimes other *Pythium* spp.

Signs and symptoms   Cuttings usually root poorly and have yellow leaves. Larger plants are stunted and wilt. Examination of the stem and roots reveals a mushy, black rot extending from the cut end into the upper portions of the stem and leaves. Roots are usually discolored and sparse, and their outer layer is easily pulled away from the stringy inner core. Root and stem rots usually occur in patches on a propagation bench where it spreads into uninfected cuttings.

Control   Control should be based on use of disease-free propagation material, sterilized potting media, and raised benches. Some isolates of *P. splendens* cause disease on many species of plants, while others are restricted to a single species. Reduce water applications to the minimum level for good rooting or crop production, and always use potting media with good aeration. Drenches with the insecticide chlorpyrifos have been shown to increase severity of Pythium root rot on Rex begonia. Preplant treatments with fungicides such as etridiazole or postplant treatment with metalaxyl or fosetyl aluminum can be very effective in control of Pythium root rot.

Selected References

Chase, A. R. 1988. Efficacy of Aliette 80WP in controlling Pythium root rot of *Begonia* and Phytophthora stem rot of *Aphelandra*. University of Florida, Central Florida Research and Education Center-Apopka, CFREC-A Research Report, RH-88-5.

Chase, A. R. 1989. 1989 fungicide trials for control of *Cylindrocladium, Helminthosporium, Pythium* and *Rhizoctonia* diseases of ornamentals. University of Florida, Central Florida Research and Education Center-Apopka, CFREC-A Research Report, RH-89-15.

Osborne, L. S., and A. R. Chase. 1987. Effects of chlorpyrifos and *Pythium splendens* on growth of Rex begonia. Plant Dis. 71:525-527.

Xanthomonas leaf spot

Figure 70

Cause   *Xanthomonas campestris* pv. *begoniae*

Signs and symptoms   Small, blisterlike spots are the initial symptoms of this disease on begonia. Spots become roughly circular, turn brown, merge, and may develop water-soaked margins. When lesions are numerous, premature leaf drop is common. Margins of lesions on Rex begonia are characteristically wavy and irregular. Many lesions form along leaf margins when infection through hydathodes occurs. On some hosts, these lesions become V-shaped with few discrete lesions forming in the leaf blade. All types of begonias have been found susceptible to this pathogen.

Control   The best way to control this disease is to use pathogen-free plants from tissue culture. All symptomatic plants should be collected and destroyed. Minimizing overhead irrigation and increasing plant spacing will also reduce disease development and spread. Using fertilizer at rates slightly higher than recommended has also been shown to reduce disease severity. High relative humidity and temperatures between 70 and 85°F favor disease development. Table 7 gives the relative resistance of some Rex begonia cultivars to this disease. Bactericides provide poor control of this disease unless all cultural control methods are utilized as well. Copper compounds and streptomycin sulfate provide some disease control, but care must be taken to use them sparingly at labeled rates to reduce the potential of pathogen resistance. Spread of the bacterium in an ebb-and-flow irrigation system was found to be minimal and not thought to be an important means of pathogen spread.

Selected References

Atmatjidou, V. P., R. P. Flynn, and H. A. J. Hoitink. 1991. Dissemination and transmission of *Xanthomonas campestris* pv. *begoniae* in an ebb and flow irrigation system. Plant Dis. 75:1261-1265.

Chase, A. R. 1992. Resistance of some Rex begonia cultivars to Xanthomonas leaf spot. Southern Nursery Digest 26(11):20-21.

Rattink, H., and H. Vruggink. 1979. A method to obtain *Xanthomonas*-free begonia plants. Med. Fac. Landbouww. Rijksuniv. Gent. 44(1):439-443.

Shaw, D. M., and A. R. Chase. 1991. Effect of fertilizer rate on susceptibility of 'Mikado' begonia to *Xanthomonas campestris* pv.

TABLE 7. Response of Rex begonia cultivars to *Xanthomonas campestris* pv. *begoniae*

Highly susceptible	Moderately susceptible	Slightly susceptible
Dew Drop	Cleopatra	Duarten
Meteor	Her Majesty	Helen Teupel
Mikado	Red Pride	Marion Louise
Phoenix Red	Tiger Kitten	Pauline
		Peace
		Red Dot
		Vesuvius

begoniae. University of Florida, Central Florida Research and Education Center-Apopka, CFREC-A Research Report, RH-91-9.

Strider, D. L. 1975. Chemical control of bacterial blight of Rieger Elatior Begonias caused by *Xanthomonas begoniae.* Plant Dis. Rep. 59:66-70.

Bromeliaceae

Aechmeas are primarily epiphytic bromeliads native to the area from Mexico to Argentina. Aechmeas are used as table or small floor plants or in mass plantings as ground covers. Plants grow best when greenhouse night temperature is maintained above 60°F (15°C) and light levels are 3,000 to 4,000 ft-c. Aechmeas require 150 ft-c or more indoors. Many other genera of bromeliads, including *Cryptanthus, Neoregelia, Nidularium, Tillandsia,* and *Vriesea,* are produced in a similar fashion. Most bromeliads are relatively disease-free once established, although small plants can be affected by foliar and soilborne fungi. Scales and root mealybugs are serious pests. In addition, when plants are produced on capillary mats, both fungus gnats and shore flies can be serious problems.

Desiccation

Figure 71

Cause   Low soil moisture or low humidity

Signs and symptoms   Plants may wilt, although this is unusual for most bromeliads because of their rigid leaves. Sometimes the leaves will curl. Patches of tan or dead tissue form on the leaf edges and tips.

Control   Make sure plants receive adequate irrigation. Keep humidities high, even when soil moisture is high, to reduce development of the leaf burn. Reducing the light level and temperature can reduce desiccation damage when humidities cannot be maintained at sufficiently high levels.

Selected Reference

Poole, R. T., and C. A. Conover. 1992. Reaction of three bromeliads to high humidity during storage. University of Florida, Central Florida Research and Education Center-Apopka, CFREC-A, RH-92-26.

Erwinia blight

Figures 72 and 73

Cause   *Erwinia* spp.

Signs and symptoms   A blackened, wet, slimy lesion generally starts at the soil line at the base of the plant and progresses into the top of the leaf and into other segments of the plants. Plants wilt, collapse, and often die.

Control   Remove and destroy infected plants as soon as they are found. Keep watering to a minimum, and avoid splashing because this can spread the bacterium to other plants. Irrigate early in the day to allow rapid drying of the foliage, thus reducing the ability of the bacterium to infect. Be sure to obtain an accurate diagnosis of the problem, because several of the diseases caused by fungi appear similar. Bactericides should not be used for control of *Erwinia* on bromeliads because they are not effective.

Fusarium stem and crown rot

Figure 74

Cause   *Fusarium* spp.

Signs and symptoms   Basal crown rot caused by a variety of species of *Fusarium* can be common when plants are established from tissue culture or unrooted offshoots. Lower leaves first appear yellow and then turn brown. A mushy stem rot is common and causes the tops to topple.

Control   Preplanting examinations of all propagative materials will usually reveal the early signs of Fusarium stem rot as small, sunken, discolored spots in the stems at the cut surface or just below the leaves. Always use new or sterilized potting media, flats, and pots to avoid spreading disease from one crop to the next. No information is available on effective fungicides for this disease on bromeliads.

Helminthosporium leaf spot

Figure 75

Cause   *Exserohilum rostratum*

Signs and symptoms   Initial lesions are pinpoint, water soaked, and chlorotic. They are circular (1 to 3 mm in diameter) to elliptical. Individual spots become sunken with brown centers and often have a narrow yellow halo. Under optimal conditions, spots merge to form large necrotic areas, causing affected leaves to collapse and hang limply on the plant. This disease can be very serious on small plants when they are transplanted, since wounding creates more infection sites.

Control   The cultural controls listed for Erwinia blight are also effective for Helminthosporium leaf spot. Fungicides that control this leaf disease include chlorothalonil, mancozeb, and zineb. Check fungicide labels for legal uses on specific species of bromeliads.

Selected Reference

Marlatt, R. B., and J. F. Knauss. 1974. A new leaf disease of *Aechmea fasciata* caused by *Helminthosporium rostratum.* Plant Dis. Rep. 58:445-448.

Pythium root rot

Figures 76 and 77

Cause   *Pythium* spp.

Signs and symptoms   Bromeliads infected with *Pythium* spp. turn a dull gray green and may wilt. Stems become rotted at the soil line, and upper portions of the plant collapse. Roots are darkened and mushy and generally sparse.

Control Use pathogen-free pots and potting media, and grow plants on raised benches. Overwatering or using potting media that drains poorly may predispose plants to attack by root-rotting fungi. Soil drenches with etridiazole or metalaxyl aid in control of stem and root rots caused by *Pythium* spp.

Rhizoctonia aerial blight

Figures 78

Cause *Rhizoctonia solani*

Signs and symptoms A mass of brownish mycelia can cover infected plants. Growth of mycelia from the potting medium onto the plant can escape notice and give the appearance that plants have been infected from an aerial source of inoculum. Close examination, however, generally reveals the presence of mycelia on stems prior to development of obvious symptoms. Under wet conditions, spots caused by *Rhizoctonia* are gray or black and mushy.

Control Cultural controls are the same as those listed for Pythium root rot. Chemical control of diseases caused by *Rhizoctonia* spp. has been investigated on many plants with a variety of fungicides. The fungicide used most widely as a soil drench to control *Rhizoctonia* is thiophanate methyl.

Slime mold

Figure 79

Cause One of many genera of Myxomycetes

Signs and symptoms Small, tan or creamy-colored structures, sometimes similar in appearance to those caused by rust diseases, can be found anywhere on the plant surface. They are easy to scrape off leaves or stem surfaces. The cause of this problem is slime mold, a saprophytic fungus that can grow on any surface, including the potting medium or pots.

Control Fungicides have not been found effective in controlling slime molds. They do not directly injure plants. Keep pots as dry as feasible for good plant growth, and grow on raised benches only.

Cactaceae and Succulents

Aloe

The genus *Aloe* is native to the arid regions of Asia and Africa. Species utilized for foliage plants are herbaceous and sold in small dish gardens and containers. Aloes are produced under 5,000 to 6,000 ft-c with a minimum night temperature of 55°F (13°C). Many serious diseases have been reported on *Aloe* spp. over the years, including rust

and Pythium root rot. Pests of *Aloe* include mealybugs and scales.

Cereus

Most *Cereus* spp. are native to the West Indies and eastern South America. They are used primarily as floor specimens and will tolerate the low light and dry conditions of some interior locations. Plants generally are produced from cuttings under 6,000 to 8,000 ft-c with a night temperature no lower than 55°F (13°C). Light levels of 100 to 150 ft-c are recommended in the interiorscape. The most serious disease is a stem rot caused by a fungus; scales are the major pest problem.

Hatiora (*Rhipsalidopsis*) and *Schlumbergera*

These epiphytic cacti are native to Brazil. Interiorscape uses are primarily confined to small table-top plants and hanging baskets. Plants are usually sold in flower, *Hatiora* (*Rhipsalidopsis*) during the spring and *Schlumbergera* at Thanksgiving or Christmas. Plants are produced under 1,500 to 3,000 ft-c with a minimum temperature of 55°F (13°C), although plants will tolerate temperatures near freezing. Indoor light levels in excess of 150 ft-c are needed to maintain attractive plants. Cactus viruses, cyst nematode, and several bacterial and fungal diseases frequently occur on these plants. The most common pests include scales and mealybugs; severe damage is also seen from fungus gnats when plants are being rooted or the potting medium is too moist.

Opuntia

Opuntia spp. are native to North and South America and are characterized by their flattened stem segments. Sizes of those used as foliage plants vary from 8 cm to 1 m, depending upon the species and intended use. Opuntias are grown under 6,000 to 8,000 ft-c and need a night temperature no lower than 55°F (13°C). Most diseases of opuntias are caused by fungi that attack the stem segments. Cactus virus X can also cause some problems. Pests include mites and scales.

Sedum

The genus *Sedum* is native to the temperate zones of the northern hemisphere. Some resemble *Crassula* spp. and can be grown under much the same conditions. Production is best under 5,000 to 6,000 ft-c with a minimum night temperature of 55°F (13°C). Leaf spots and stem rots on *Sedum* can be caused by both bacteria and a variety of fungi; mealybugs and scales are troublesome pests.

Senecio

These plants are generally vining and produced in hanging baskets under the same conditions used for many succulents and cacti. Senecio should be grown under 4,000 to 6,000 ft-c with a minimum night temperature of 55°F (13°C). Many fungi cause diseases of this plant, and scales and mealybugs are the most common pests.

Anthracnose

Figure 80

Cause *Colletotrichum gloeosporioides*

Signs and symptoms Leaf and stem spots appear as soft areas with sunken centers. These spots can turn black or tan depending upon the plant species. The spores of the fungus are orange and are easily transferred by splashing irrigation water or rainfall. Plants that are susceptible include *Crassula, Graptopetatum, Sedum,* and *Sempervivum* spp.

Control This disease is rarely a problem on mature or well-rooted plants, since wounds are needed for infection. For this reason, special care in protecting rooting cuttings can generally control the problem completely. Iprodione is effective on anthracnose and labeled for *Crassula.* Check fungicide labels for possible products to use on other succulents susceptible to anthracnose.

Selected Reference

Chase, A. R. 1983. Leaf rot of four species of Crassulaceae caused by *Colletotrichum gloeosporioides.* Plant Pathol. 32:351-352.

Botrytis blight

Figure 81

Cause *Botrytis cinerea*

Signs and symptoms Botrytis blight usually appears on portions of cuttings in contact with the potting medium or in the center of the plant where the humidity is highest. The water-soaked spots enlarge rapidly to encompass a large portion of the plant. When night conditions are cool, day conditions warm, and moisture conditions high, the pathogen readily sporulates on infected plant parts, covering them with grayish brown dusty masses of conidia (spores).

Control In the greenhouse, it is particularly important to control Botrytis blight on most plants during the winter months. Many methods that increase foliar drying and reduce moisture condensation on plants during the night reduce the need for fungicide application. Chlorothalonil and vinclozolin are effective in controlling Botrytis blight on many plants. This disease is rarely found in commercial production of succulents and cacti.

Cactus cyst nematode

Figure 82

Cause *Cactodera cacti* (= *Heterodera cacti*)

Signs and symptoms Heavily infected plants are stunted, foliage turns red brown, and wilting is common. Examina-

tion of the roots reveals tiny, round cysts, which may be white when immature and turn golden to medium brown when mature. The cyst is the female nematode and is usually attached to the roots.

Control Always raise plants above the ground and use nematode-free potting medium, pots, and plant materials. Many nematicides are effective in controlling this nematode problem. Research has shown that ethoprop provides excellent control but is severely phytotoxic, as is fensulfothion. Carbofuran and oxamyl provide good nematode control and allow normal foliage growth. Be sure to consult current nematicide labels for legal uses on specific cacti.

Selected References

Hamlen, R. A. 1975. Evaluation of nematicides for control of *Heterodera cacti* affecting *Zygocactus truncatus.* Plant Dis. Rep. 59: 636-637.

Langdon, K. R., and R. P. Esser. 1969. Cactus cyst nematode, *Heterodera cacti,* in Florida, with a host list. Plant Dis. Rep. 53:123-125.

O'Bannon, J. H., and R. P. Esser. 1970. Control of *Heterodera cacti* infecting *Zygocactus truncatus.* Plant Dis. Rep. 54:692-694.

Cercospora leaf spot

Figure 83

Cause *Cercospora* sp.

Signs and symptoms Tiny, slightly raised, red or dark green spots appear on the lower leaf surfaces. Spots enlarge slowly and eventually are visible on the tops of leaves as indistinct or blotchy yellow and brown areas.

Control Use pathogen-free plants as cutting sources, minimize leaf wetting, and remove plants that have spots after frequent scouting. No information is available on fungicide control of this disease on succulents.

Dichotomophthora rot

Figure 84

Cause *Dichotomophthora indica*

Signs and symptoms Spots are generally tan and may be dry and appear sunken. The black spores of this fungal pathogen are sometimes found within the spots, and they are easily spread by splashing water. There are only a few cacti reported as hosts of this pathogen (e.g., *Opuntia* and *Gymnocalycium*); however, many more are probably susceptible.

Control See controls listed for Helminthosporium blight below.

Selected Reference

Pfeiffer, C. M., J. E. Wheeler, D. A. Bach, and R. L. Gilbertson. 1989. First report of *Dichotomophthora indica* as a pathogen of *Myritillocactus geometrizans* and *Gymnocalycium mihanovichii* var. *friedrichii* in Arizona. Plant Dis. 73:81.

Erwinia blight

Figures 85 and 86

Cause   *Erwinia* spp.

Signs and symptoms   A blackened, wet, slimy spot generally starts at the soil line at the base of the plant and progresses into the top of the cladophyll or into the upper portions of the plant. Plants wilt, collapse, and often die. Because the fungus produces macerating enzymes, infected plants become very mushy and disintegrate, especially during the warm months of the year.

Control   Remove and destroy infected plants as soon as they are found. Keep watering to a minimum, and avoid splashing, since this can spread the bacterium to other plants. Irrigate early in the day to allow rapid drying of the foliage, which will reduce the ability of the bacterium to infect. Be sure to obtain an accurate diagnosis of the problem, since several of the diseases caused by fungi are similar in appearance. Use of bactericides on plants infected with *Erwinia* spp. is rarely effective, and none are labeled for bacterial soft rot on cacti and succulents. Most cacti and succulents will die if infected with *Erwinia* spp., so prevention is the key to control. Table 8 lists the levels of resistance of *Schlumbergera truncata* cultivars to this disease.

Selected References

Chase, A. R., and J. M. F. Yuen. 1993. Susceptibility of *Schlumbergera truncata* cultivars to four plant pathogens. J. Environ. Hort. 11:14-16.

Jump, J. A., F. Mittermeyer, and K. S. Price. 1983. The cause and control of the soft rot of Lithops. Cactus & Succulent Journal (U.S.) 55:65-68.

Suslow, T., and A. H. McCain. 1979. Etiology, host range, and control of a soft rot bacterium from cactus. (Abstr.) Phytopathology 69:921.

Ethylene damage

Figure 87

Cause   Ethylene gas

Signs and symptoms   Plants develop watery dark spots on leaves and stems. The spots can look the same as those caused by some fungi and bacteria. Leaf drop is common.

Control   These symptoms appear during shipping when plants are exposed to ethylene. Since fruits and vegetables generate this gas, avoid shipping with produce. Storing plants under conditions of high humidity and temperature can also promote ethylene damage. Special padding that absorbs ethylene gas is available for lining shipping containers, and it may prove effective.

Fusarium rot

Figures 88, 89, and 90

Cause   *Fusarium oxysporum*

Signs and symptoms   An infection appears at the cladophyll border or in plant centers. Spots are generally tan and may be dry at times and appear sunken. The orange spores of the pathogen form in the lesions, and they are easily spread by water or air since they are light weight. Abscission of cladophylls above the affected portion can occur when conditions are wet and warm. In addition, when moisture is high, the mycelium of the fungus can develop extensively and cover the entire plant. Nearly all cacti and succulents are susceptible to *Fusarium* diseases. Many cacti infected with *Fusarium* spp. collapse completely, and their centers become engulfed in the white mycelium of the fungus.

Control   Use the same cultural controls as listed for soft rot. Thiophanate methyl and mancozeb are effective in controlling *Fusarium* diseases. Reduction of water applica-

TABLE 8. Susceptibility of some *Schlumbergera truncata* (holiday cacti) cultivars to four pathogens

Cultivar	*Erwinia*	*Drechslera*	*Fusarium*	*Phytophthora*
Bridgeport	Moderate	Moderate	Low	Moderate
Cambridge	High	Moderate	Variable	High
Christmas Charm	Variable	Moderate	Moderate	Moderate
Christmas Fantasy	Low to moderate	Moderate	Variable	Variable
Christmas Flame	Variable	Moderate	High	Moderate
Christmas Magic II	Low	Moderate	Low	Low
Gold Charm	High	Variable	High	High
Holiday Splendor	Variable	Moderate	Low	High
Kris Kringle	Low	High	Moderate	Variable
Kris Kringle II	Low to moderate	High	Variable	Variable
Lavender Doll	Variable	Low to moderate	Low	Variable
Lavender Doll II	Low	Variable	Variable	Moderate to high
Peach Parfait	Variable	High	Low	High
Red Radiance	Variable	Moderate	Moderate	Variable
Sanibel	Variable	Variable	Low	Moderate
Santa Cruz	Variable	Low to moderate	Moderate	Moderate
Sleigh Bells	Variable	Moderate	Low	Variable
Twilight Tangerine	High	Moderate	Moderate	Moderate
White Christmas	Low	Low to moderate	Moderate	Moderate
Windsor	Variable	Moderate	Variable	Variable

tions greatly inhibits spread of the pathogen to aerial portions of the plants, thus reducing disease severity. Table 8 lists the levels of resistance of *Schlumbergera truncata* cultivars to this disease.

Selected References

Chase, A. R., and J. M. F. Yuen. 1993. Susceptibility of *Schlumbergera truncata* cultivars to four plant pathogens. J. Environ. Hort. 11:14-16.

Mitchell, J. K. 1987. Control of basal stem and root rot of Christmas and Easter cacti caused by *Fusarium oxysporum*. Plant Dis. 71:1018-1020.

Moorman, G. W., and R. A. Klemmer. 1980. *Fusarium oxysporum* causes basal rot stem of *Zygocactus truncatus*. Plant Dis. 64: 1118-1119.

Raabe, R. D., J. H. Hurlimann, and B. Bruckner. 1975. A root and stem rot complex of Christmas cactus and its control. Cal. Pl. Path. 28:1-2.

Helminthosporium rot

Figures 91, 92, and 93

Cause   *Drechslera cactivora* (= *Helminthosporium cactivorum*)

Signs and symptoms   Blackened, sunken lesions from 1 to 15 mm across form on all aboveground portions of the plant. The spots on holiday cacti are generally circular and can also occur below ground. Cladophyll abscission is common on plants, even when infection appears light. The black spores of the fungus form in the spots, giving them a fuzzy appearance. The disease was first described in the mid-1950s on *Cereus* and remains a problem for some producers today. A rapid rot of the cotyledons of young cacti is one of the first symptoms. Older plants become rotted where spines have broken or the stem has been punctured. Most plants blacken and may be mushy or dryish, and the affected portion collapses. *Hatiora* (*Rhipsalidopsis*) is very susceptible to Drechslera leaf spot and *Schlumbergera* is moderately susceptible.

Control   Use the same cultural controls as listed for soft rot. Disease is most severe at temperatures between 75 and 91°F (24 and 36°C). In addition, sprays and drenches of chlorothalonil are very effective in controlling the disease but should be used with caution, since it has been shown on occasion to cause slight chlorosis. Table 8 lists the levels of resistance of *Schlumbergera truncata* cultivars to this disease.

Selected References

Chase, A. R. 1982. Stem rot and shattering of Easter cactus caused by *Drechslera cactivora*. Plant Dis. 66:602-603.

Chase, A. R., and J. M. F. Yuen. 1993. Susceptibility of *Schlumbergera truncata* cultivars to four plant pathogens. J. Environ. Hort. 11:14-16.

Durbin, R. D., L. H. Dais, and K. F. Baker. 1955. A Helminthosporium stem rot of cacti. Phytopathology 45:509-512.

Phytophthora stem rot

Figure 94

Cause   *Phytophthora parasitica* and other *Phytophthora* spp.

Signs and symptoms   Cacti infected with *Phytophthora* spp. turn a dull gray green and may wilt. Stems become rotted at the soil line, and upper portions of the plant collapse. Cladophyll abscission may occur. Roots are darkened and mushy and generally sparse. Most cacti and succulents can be infected with *Phytophthora* spp. if they are overwatered or planted in potting medium that drains poorly.

Control   Use pathogen-free pots and potting media and grow plants on raised benches. Overwatering plants may predispose them to attack by root-rotting fungi. Soil drenches with etridiazole or metalaxyl aid in control. Table 8 lists the levels of resistance of *Schlumbergera truncata* cultivars to this disease.

Selected References

Alfieri, S. A., Jr., and J. W. Miller. 1971. Basal stem and root rot of Christmas cactus caused by *Phytophthora parasitica*. Phytopathology 61:804-806.

Averre, C. W, III, and J. E. Reynolds. 1964. Phytophthora root and stem rot of *Aloe*. Proc. Fla. State Hortic. Soc. 77:438-440.

Chase, A. R., and J. M. F. Yuen. 1993. Susceptibility of *Schlumbergera truncata* cultivars to four plant pathogens. J. Environ. Hort. 11:14-16.

Knauss, J. F. 1975. Control of basal stem and root rot of Christmas cactus caused by *Pythium aphanidermatum* and *Phytophthora parasitica*. Proc. Fla. State Hortic. Soc. 88:567-571.

Raabe, R. D., J. H. Hurlimann, and B. Bruckner. 1975. A root and stem rot complex of Christmas cactus and its control. Cal. Pl. Path. 28:1-2.

Pythium root rot

Figure 95

Cause   *Pythium* spp.

Signs and symptoms   Foliage of plants infected with *Pythium* spp. turns a dull gray green and may wilt. Stems become rotted at the soil line, and upper portions of plants collapse. Cladophyll abscission may occur. Roots are darkened and mushy and generally sparse.

Control   Cultural and chemical controls are the same as those listed for Phytophthora stem rot.

Selected References

Baker, K. F., and K. Cummings. 1943. Control of Pithium root rot of *Aloe variegata* by hot-water treatment. Phytopathology 33:736-738.

Knauss, J. F. 1975. Control of basal stem and root rot of Christmas cactus caused by *Pythium aphanidermatum* and *Phytophthora parasitica*. Proc. Fla. State Hortic. Soc. 88:567-571.

Caladium

Caladiums are native to the tropical Americas and the West Indies. They are produced for use in single pots (15 cm or larger) as well as for ground plantings, especially in the southeastern states. Most caladiums are produced in partial shade and should be used under similar conditions. Caladium diseases are caused primarily by soilborne pests such as fungi and nematodes. Dasheen mosaic virus and Xanthomonas blight also cause problems under some conditions.

Aspergillus tuber rot

Figure 96

Cause *Aspergillus niger*

Signs and symptoms The pathogen attacks primarily wounded or damaged tubers. Since the process of digging caladium tubers often wounds them, *Aspergillus* can sometimes be a problem. Small, sunken, brown, rotted areas form, and under warm conditions, the fungus sporulates, giving the tuber a fuzzy black covering. The black sporangia are slightly larger than pepper grains and are easy to see with the naked eye. It is not uncommon for other pathogens, such as *Fusarium* spp., to be present as well. In some cases, the tubers can still generate vigorous plants, but in others the tuber is so weakened that it cannot produce a good plant.

Control Prestorage dips or dusts of tubers with fungicides aid in control of this tuber rot. Hot water treatments (about 50°C [122°F]) have been effective in controlling other pathogens and pests that are carried on caladium tubers. Be sure to test the effect of temperature on the viability of the caladium before attempting this treatment on a broad scale. Store tubers at 70°F (21°C), since higher temperatures allow pathogens to develop and lower temperatures can result in damage to the tuber, thus weakening it defense against tuber-rotting fungi.

Botrytis blight

Figure 97

Cause *Botrytis cinerea*

Signs and symptoms Tan to brown spots can form on leaves and petioles. Under severe conditions, all of the leaves may be infected, leading to complete melting down of the plant. Botrytis blight occurs primarily during the cool, wet, cloudy times of the year, especially during the winter or early spring.

Control Reducing the relative humidity by increasing spacing between plants, watering by a method that does not wet the foliage, and venting and heating the greenhouse late in the day will each reduce the conditions that favor development and spread of this disease. Chlorothalonil and iprodione provide excellent control of Botrytis blight on many crops.

Cold damage

Figure 98

Cause Exposure to temperatures below 65°F (18°C)

Signs and symptoms Dead areas appear on leaf margins and tips and sometimes in centers. They are tan and papery and tear easily.

Control Do not expose caladiums to temperatures below 65°F (18°C). Protect landscape beds of caladiums by covering them with sheets if a cold snap is forecast.

Dasheen mosaic

Figure 99

Cause Dasheen mosaic virus (DMV)

Signs and symptoms Dasheen mosaic virus can be a severe problem when infected tubers are used to produce plants. The resulting crops will be infected with the virus, and tuber size and numbers will be reduced compared with virus-free crops. Symptoms include mosaic, leaf distortion, and stunting and appear periodically during the year, depending upon the environmental conditions.

Control DMV is sometimes spread by aphids, and they should be strictly controlled. It is very important to use pathogen-free stock whenever possible, since the symptoms of DMV are not always expressed. There are no chemical treatments for this viral disease. Other hosts of this virus include dieffenbachia, philodendron, taro, and anthurium.

Selected References

Hartman, R. D. 1974. Dasheen mosaic virus and other phytopathogens eliminated from caladium, taro, and cocoyam by culture of shoot tips. Phytopathology 64:237-240.

Hartman, R. D., and F. W. Zettler. 1974. Effects of dasheen mosaic virus on yields of caladium, dieffenbachia and philodendron. Phytopathology 64:768.

Zettler, F. W., and R. D. Hartman. 1987. Dasheen mosaic virus as a pathogen of cultivated aroids and control of the virus by tissue culture. Plant Dis. 71:958-963.

Fusarium tuber rot

Figures 100 and 101

Cause *Fusarium solani*

Signs and symptoms Fusarium tuber rot occurs initially on new tubers as small, discolored, sunken areas. In advanced infections, the tubers are completely rotted and have a dry, chalky texture. The pathogen may form spores, which are yellow to tan. If tubers contaminated with *Fusarium* are planted, the resulting plants will be stunted and of poor quality.

Control Use the same controls as mentioned for Aspergillus tuber rot. Fungicides such as chlorothalonil, thiophanate methyl, and mancozeb provide excellent control of *Fusarium* diseases.

Selected Reference

Knauss, J. F. 1975. Description and control of Fusarium tuber rot of *Caladium*. Plant Dis. Rep. 59:975-979.

Rhizoctonia aerial blight

Figure 102

Cause   *Rhizoctonia* spp.

Signs and symptoms   Rhizoctonia aerial blight occurs primarily during the summer or warmer months. Disease development can occur in less than a week, so plants should be checked carefully and frequently. Brown, irregularly shaped spots form anywhere on the foliage but most commonly within the crown of the plant, which is often wet. Tuber and petiole rot also occur. The disease spreads rapidly, and the entire plant can become covered with the brown weblike mycelium of the pathogen.

Control   A pathogen-free potting medium is the first step in controlling all soilborne pathogens. Plants should be produced from pathogen-free stock and grown in new or sterilized pots on raised benches. Since this pathogen inhabits the soil, both the roots and the foliage of the plants must be treated with a fungicide to provide optimal disease control. A combination drench-spray will best accomplish this. PCNB (pentachloronitrobenzene) and thiophanate methyl may be effective for control of tuber rots on caladiums.

Selected Reference

Chase, A. R. 1991. Characterization of *Rhizoctonia* species isolated from ornamentals in Florida. Plant Dis. 75:234-238.

Root-knot nematode

Figure 103

Cause   *Meloidogyne* spp.

Signs and symptoms   Galls occur on roots, and the root system may be drastically reduced. Plant stunting and wilting occur when infestations are severe.

Control   Use steam-treated soil or soilless potting media, and grow plants on raised benches if possible. Hot water treatment of tubers at 122°F (50°C) for 30 min effectively reduces nematode infestations of tubers. Oxamyl will aid in control, but be sure to check labels for this plant and legal use methods.

Selected References

Overman, A. J., and B. K. Harbaugh. 1983. Soil fumigation increases caladium tuber production in sandy soil. Proc. Fla. State Hortic. Soc. 96:248-250.

Rhoades, H. L. 1970. A comparison of chemical treatments with hot water for control of root-knot nematodes in *Caladium* tubers. Plant Dis. Rep. 54:411-413.

Rhoades, H. L., and R. A. Hamlen. 1975. Response of root-knot-infected caladiums, with and without hot water treatment, to foliar applications of oxamyl and phenamiphos. Plant Dis. Rep. 59:91-93.

Southern blight

Figure 104

Cause   *Sclerotium rolfsii*

Signs and symptoms   The pathogen attacks all portions of the plants but is most easily seen at stem bases. Initial stem symptoms are confined to water-soaked areas near the soil line. White, relatively coarse mycelia grow in a fanlike pattern on the surfaces of leaves or the potting medium. The sclerotia of the fungus form in these mycelia and are initially white but turn dark brown when mature and are the size of a mustard seed. A cutting rot can occur during warm months.

Control   All infected plants and the pots they are in should be removed from the growing area and destroyed as soon as the disease is found. Many other plant species are susceptible to this pathogen. They should be considered potential sources of infection and should be protected from an infected caladium as well. PCNB is effective for this disease on caladium.

Sunscald

Figure 105

Cause   High light levels

Signs and symptoms   Leaves rapidly develop tan to white, bleached patches. The spots can appear overnight after exposure of plants to light much brighter than that to which they are acclimated.

Control   Never move a plant from a low-light or shaded condition into direct or very bright sunlight. If plants must be moved to higher light, make the move in increments from low to medium to high light over a period of weeks to allow the leaves to adjust to the higher light levels.

Xanthomonas leaf spot

Figure 106

Cause   *Xanthomonas campestris* pv. *dieffenbachiae*

Signs and symptoms   Leaf spots on caladium start as tiny, water-soaked areas that can rapidly enlarge to 3 mm or more. They tend to form on leaf margins, but when the bacterium infects a major leaf vein, the entire leaf can collapse. Spots are frequently very black and surrounded by a bright yellow halo. Most other plants in the Aroid family such as *Aglaonema, Anthurium, Dieffenbachia,* and *Syngonium* spp. are also hosts of this pathogen.

Control   Eliminate all stock plants that have Xanthomonas leaf spot. All bacterial diseases are very difficult to control unless plants are grown without overhead watering or exposure to rainfall. Bactericides such as copper-containing compounds may be somewhat effective if used on a preventative and regular basis, but none are legal for use on caladiums at this time.

Selected Reference

Chase, A. R., R. E. Stall, N. C. Hodge, and J. B. Jones. 1992. Characterization of *Xanthomonas campestris* strains from aroids using physiological, pathological, and fatty acid analyses. Phytopathology 82:754-759.

Calathea

Calatheas are herbaceous plants native to tropical America. Most are used in dish gardens or terrariums and as desk-top specimen plants. They are relatively tolerant of indoor conditions when relative humidities are maintained above 25% and light levels of 100 ft-c or more are supplied. Production light levels of 1,000 to 2,000 ft-c are recommended, with a minimum greenhouse temperature of 55°F (13°C). Calatheas are subject to several fungal leaf diseases as well as fungal root and stem rots. Most recently, bacterial infections have been reported on Calatheas and are causing severe losses in production. Tarsonemid and spider mites, mealybugs, scales, and lepidopterous larvae are common pests. Spider mites are very common pests and seriously limit the longevity of calatheas once indoors.

Alternaria leaf spot

Figure 107

Cause   *Alternaria alternata*

Signs and symptoms   Alternaria leaf spot of *Calathea* spp. is characterized by small (less than 1 mm in diameter) spots, which are initially water soaked. These spots turn reddish brown, may reach 2 mm in diameter, and are roughly circular. Spots generally do not merge. *Calathea bella*, *C. insignis* (rattlesnake plant), and *C. picturata* 'Argentea' are susceptible to this disease; *C. bella* is very susceptible.

Control   Alternaria leaf spot can be controlled by eliminating overhead irrigation and exposure to rainfall and by the use of pathogen-free plants. Routine scouting and removal of infected plants also helps reduce disease spread. Chlorothalonil, mancozeb, and iprodione are effective for controlling this disease on calatheas.

Selected Reference

Chase, A. R. 1982. Alternaria leaf spot of *Calathea* spp. Plant Dis. 66:953-954.

Burrowing nematode

Figure 108

Cause   *Radopholus similis*

Signs and symptoms   Symptoms include lesions on roots and root rot, which lead to reduced plant vigor and poor growth. Both roots and tops appear stunted.

Control   Use new or sterilized potting media and pots, grow plants on raised benches, and always use nematode-free cuttings. Control achieved with aldicarb was reduced in a mix of peat, pine bark, and cypress shavings (2:1:1, by volume) compared with that achieved in a 3:1 mix of peat and sand. In contrast, ethoprop and oxamyl worked equally well against burrowing nematode on *C. makoyana* in the two potting media tested.

Selected Reference

Hamlen, R. A., and C. A. Conover. 1977. Response of *Radopholus similis*-infected *Calathea* spp., container-grown in two soil media, to applications of nematicides. Plant Dis. Rep. 61:532-535.

Cucumber mosaic

Figure 109

Cause   Cucumber mosaic virus (CMV)

Signs and symptoms   Leaves may be slightly distorted and reduced in size, but the most obvious symptom of CMV infection is the bright yellow patterns formed on the leaves. These patterns are generally jagged and alternate with the normal coloration of the affected leaf. CMV causes minor symptoms on most calatheas.

Control   There is no evidence that the damage caused is other than aesthetic. The only recommended control is removal of the plant material with these symptoms. Propagation of infected material will transfer the virus to new plants.

Desiccation

Figure 110

Cause   Lack of sufficient water

Signs and symptoms   Leaves sometimes wilt, but often the first sign of desiccation is loss of color and browning of leaf margins and tips.

Control   Leaf burn from lack of water can be caused by poor irrigation practices; root pathogens or nematodes, which can reduce the roots' ability to absorb water; and even spider mites, which can cause localized desiccation by extensive feeding. Be sure to determine the cause of the desiccation before attempting a control method. Lifting pots to determine whether water is being held in the potting medium (heavier ones have more water) and then examining roots to make sure they are not affected by pathogens and nematodes are critical.

Fluoride toxicity

Figure 111

Cause   Excess fluoride

Signs and symptoms Dead spots form near leaf margins and may be discrete or have blurred margins.

Control Maintain the pH of the potting medium at about 6.5 to reduce the availability of soluble fluoride (lower pH levels make fluoride more available). Amending potting media with dolomite and avoiding the use of perlite and superphosphate are also recommended. Certain peats and sometimes irrigation water are sources of fluoride as well. *Calathea makoyana* is more tolerant of fluoride than *C. insignis*.

Selected Reference

Conover, C. A., and R. T. Poole. 1984. Relationships of dolomite and superphosphate to production of *Calathea*. University of Florida, Agricultural Research and Education Center-Apopka, AREC-A Research Report, RH-84-16.

Foliar nematode

Figure 112

Cause *Aphelenchoides besseyi*

Signs and symptoms Leaf spots begin near the midvein on lower leaves and extend to the margin. They are usually rectangular.

Control Infestation of *Calathea* occurs through movement of nematodes from the soil to the lower leaves. If possible, grow these plants in containers on benches away from the source of infestation. Many other plants are hosts of foliar nematode including begonias, hosta, ferns, and geraniums. Be sure to check for nematode symptoms on all plants in your landscape or greenhouse.

Fusarium wilt

Figure 113

Cause *Fusarium oxysporum*

Signs and symptoms This disease can be common on calatheas propagated from cuttings. Cutting bases may rot, or lower leaves can wilt and turn yellow. Larger plants show signs of wilting and yellowing of lower leaves, and the vascular system is brownish. Plants eventually become stunted.

Control Drench treatments with thiophanate methyl compounds may aid in reducing disease spread, but complete control can be achieved only if the infected planting material is destroyed. Be sure to obtain an accurate diagnosis, since symptoms of nematode infestation and root rot caused by soilborne fungi are similar in appearance.

Helminthosporium leaf spot

Figure 114

Cause *Drechslera setariae*

Signs and symptoms Most cultivars of calathea are susceptible to this pathogen. Spots first appear as tiny, water-soaked areas that turn yellow and finally brown and dry. Spots are normally large (1 cm wide). In severe cases, spots merge and form large, irregularly shaped areas, which are tan with a yellow edge.

Control Minimizing the period of time leaves are wet can dramatically reduce disease severity. Eliminate overhead watering, or at least apply water early in the day to allow rapid drying of foliage. Plants that are watered late in the afternoon may remain wet for the entire night, promoting disease. Disease is most severe at temperatures between 60 and 65°F (15 to 18°C). Table 9 shows *Calathea* cultivar resistance to Helminthosporium leaf spot. Mancozeb has been effective for control of this disease experimentally, but labels should be checked for legal use on calathea.

Selected References

Chase, A. R. 1986. Fertilizer level does not affect severity of Helminthosporium leaf spot of calatheas. University of Florida, Agricultural Research and Education Center-Apopka, AREC-A Research Report, RH-86-19.

Chase, A. R. 1987. Susceptibility of *Calathea* species and cultivars to *Bipolaris setariae*. J. Environ. Hort. 5:29-30.

Chase, A. R. 1989. 1989 fungicide trials for control of *Cylindrocladium, Helminthosporium, Pythium* and *Rhizoctonia* diseases of ornamentals. University of Florida, Central Florida Research and Education Center-Apopka, CFREC-A Research Report, RH-89-15.

Chase, A. R., and T. K. Broschat. 1991. Effect of potassium level on severity of Drechslera leaf spot of *Calathea picturata* 'Vandenheckei'. J. Environ. Hort. 9:101-102.

Simone, G. W., and D. D. Brunk. 1983. New leaf spot disease of *Calathea* and *Maranta* spp. incited by *Drechslera setariae*. Plant Dis. 67:1160-1161.

Potassium deficiency

Figure 115

Cause Low potassium

Signs and symptoms Spots form in leaf centers in a random pattern and frequently look like Alternaria or Hel-

TABLE 9. Response of some *Calathea* species and cultivars to Helminthosporium leaf spot (*Bipolaris setariae*) and Pseudomonas blight

Calathea sp. and cultivar	*Helminthosporium*	*Pseudomonas*
argentea		
'Silver Portrait'	Very high	Unknown
'Vandenheckei'	High	Unknown
concinna	Unknown	Slight
insignis 'Rattlesnake'	High	Unknown
louisae 'Green Feather'	Very low	Unknown
majestica	Unknown	Low to medium
makoyana 'Peacock'	Medium	Immune
ornata	Unknown	Very high
roseo-picta	Low	Low to medium
vittata	Unknown	Immune
wilson-princep	Unknown	High

minthosporium leaf spot. Spots are reddish and circular when mature.

Control   Use fertilizer with an adequate source of potassium. Submit leaves with these symptoms for disease diagnosis as well as nutrient analysis, since symptoms are so similar.

Selected References

Chase, A. R., and R. T. Poole. 1988. Nutritional response of five Calatheas. Proc. Fla. State Hortic. Soc. 101:323-325.
Conover, C. A., and R. T. Poole. 1984. Growth of *Calathea makoyana* as influenced by media, fertilizer and irrigation. University of Florida, Agricultural Research and Education Center-Apopka, AREC-A Research Report, RH-84-19.

Pseudomonas leaf spot

Figure 116

Cause   *Pseudomonas cichorii*

Signs and symptoms   Spots are water soaked and turn dark green to black. They may have a yellow edge, but this is not common. *Calathea roseo-picta* and cultivar Vandenheckei are most susceptible to this bacterial pathogen, and spots may reach 2.5 cm in diameter. There are rarely more than two spots on a leaf, although loss of the leaf often occurs if it is infected before it expands completely.

Control   Avoid overhead watering as much as possible to reduce conditions for infection and spread of the pathogen. Preventive applications of a copper bactericide are rarely effective and are not labeled for use on calathea at this time.

Selected References

Chase, A. R. 1993. Bacterial leaf spot and blight of calatheas. University of Florida, Central Florida Research and Education Center-Apopka, CFREC-A Research Report, RH-93-11.
Wick, R. L., and R. Shrier. 1990. *Pseudomonas cichorii* leaf spot of *Calathea picturata* 'Argentea'. (Abstr.) Phytopathology 80:124.

Pseudomonas blight

Figure 117

Cause   *Pseudomonas* sp.

Signs and symptoms   This bacterial disease has become a problem for calathea growers during the past 2 years. It is caused by a nonfluorescent pseudomonad that has not yet been identified to species. Symptoms start as watersoaked areas along the leaf veins and are especially visible on new leaves as they open up. The spots appear clear and coalesce readily. When completely mature, the spots are tan to brown and papery. *C. roseo-lineata* is very susceptible to this pathogen, and severe symptoms develop even when drip irrigation is used to keep leaves dry.

Control   Infected plants should be destroyed. Although tissue-culture plants may be free of the pathogen when purchased, they are as easily infected as cuttings. Other controls mentioned for *P. cichorii* can be used. Table 9 gives responses of some *Calathea* species and cultivars to this disease.

Selected References

Chase, A. R. 1993. Bacterial leaf spot and blight of calatheas. University of Florida, Central Florida Research and Education Center-Apopka, CFREC-A Research Report, RH-93-11.
Leahy, R. M. 1991. Bacterial leaf spot of *Calathea* spp. Fla. Dept. Agric. & Cons. Serv. Plant Pathol. Circ. No. 345.

Chlorophytum

Although there are a great many diseases listed in *Diseases and Disorders of Plants in Florida for Chlorophytum comosum*, little, if any, information has been published on diseases of this foliage plant. Fungal leaf spots caused by *Alternaria, Cercospora, Fusarium,* and *Phyllosticta* spp. have been listed as well as fungal root diseases caused by *Pythium splendens, Rhizoctonia solani,* and *Sclerotium rolfsii* (Southern blight). Bacterial pathogens are also listed and include *Erwinia carotovora* and *Pseudomonas cichorii*. These organisms are known to cause serious diseases of other foliage crops, but their roles in diseases of *C. comosum* have not been established. In addition, during the past 20 years spent studying foliage plant diseases at the CFREC-Apopka, we have not encountered any serious diseases of this plant.

Tipburn

Figure 118

Cause   Excess boron or fluoride

Signs and symptoms   Leaves have necrotic tips or necrotic areas within the white areas. Sometimes chlorotic areas appear between the necrotic spots and green areas. Tipburn is most common. When the cause is fluoride toxicity, there is a reddish border between the necrotic tissue and healthy tissue. This border is tan to gray when the cause is boron toxicity.

Control   Use irrigation water free of boron and fluoride, and select medium components without these contaminants. Be sure fertilizers are free of fluoride when possible, and use no more than the equivalent of 0.5 lb of boron per acre per year (0.43 g/1,000 ft^2/month). Maintain a medium pH of 6.0 to 6.5 to reduce availability of boron and fluoride.

Selected Reference

Ben-Jaacov, J., R. T. Poole, and C. A. Conover. 1984. Tipburn of *Chlorophytum comosum* 'Vittatum'. HortScience 19:445-447.

Cissus (Grape ivy)

Cissus are popular foliage plants used in hanging baskets and as ground covers in mass plantings. These plants are native to the tropics and subtropics, and vining forms are commonly used as foliage plants. Cissus should be produced under 1,200 to 2,000 ft-c, and 75 to 150 ft-c should be provided in the interiorscape. Many fungi cause leaf, stem, and root diseases of cissus. The most common pests include spider and tarsonemid mites, scales, lepidopterous larvae, and mealybugs.

Anthracnose

Figures 119 and 120

Cause   *Colletotrichum* sp.

Signs and symptoms   Anthracnose of both grape ivy and kangaroo vine often occurs during the rooting process. Many leaves can be affected, and cuttings are lost because of leaf rot. Single spots appear anywhere on the leaf and are water soaked and roughly round and sometimes contain the fruiting bodies of the pathogen in concentric rings. The fruiting bodies are black and the size of pepper grains. When spots dry out, they turn tan to gray and are papery. Large, well-established plants are also susceptible to *Colletotrichum* sp. Under conditions of high moisture and high disease pressure, many small (1 to 2 mm), angular (bordered by leaf veins) spots form. The youngest leaves are the most susceptible.

Control   Use only disease-free stock plants for cuttings, since infected stock plants rarely give rise to healthy new plants. Disease is reduced by minimizing the amount of water applied to leaves and increasing plant spacing, which enhances rapid drying of the foliage. Fungicides such as mancozeb and thiophanate methyl are effective in controlling anthracnose.

Selected Reference

Humphreys-Jones, D. R. 1976. Leaf blotch (*Glomerella cingulata* (Stonem.) Spauld. & Schrenck) on *Cissus antarctica* Vent. Plant Pathol. 25:115.

Bendiocarb phytotoxicity

Figure 121

Cause   Soil drenches of bendiocarb

Signs and symptoms   Leaves are curled downward and sometimes have brown spots.

Control   Do not apply any pesticide to the potting medium that is labeled only for foliar application. Using pesticides in ways inconsistent with their labels is illegal and can cause phytotoxicity.

Selected Reference

Osborne, L. S., and A. R. Chase. 1986. Phytotoxicity evaluations of Dycarb on selected foliage plants. University of Florida, Agricultural Research and Education Center, AREC-Apopka, Research Report, RH-86-12.

Botrytis blight

Figure 122

Cause   *Botrytis cinerea*

Signs and symptoms   *Botrytis* infections are typified by large, grayish areas on leaf margins and sometimes leaf centers. Leaves in plant centers are commonly infected, since these areas remain wet longest. The dusty gray to tan spores of the pathogen form all over the tissue and can easily be seen with a 10× hand lens. Botrytis blight is most serious during the relatively cool and dark winter months.

Control   Many plants other than *Cissus* spp. can be affected by Botrytis blight, and control measures should be extended to all susceptible crops. Examples of these crops include lipstick vine, hoya, African violet, and English ivy. Reducing moisture levels around the foliage by limiting water applications and increasing air movement are recommended cultural controls for this and many other foliar diseases. Iprodione and vinclozolin are very effective for *Botrytis* control on many plants.

Downy mildew

Figure 123

Cause   *Plasmopara viticola*

Signs and symptoms   This disease is a problem primarily during the cool, cloudy, winter months. Yellowish to brown blotches appear on upper surfaces of leaves, frequently along the leaf margins. Under moist, cool conditions, leaves may turn yellow and droop. A light brown to purplish fungal growth may appear on the lower surfaces of the leaves. Under drier conditions, leaf spots appear as brown, burned areas.

Control   Downy mildew occurs during the cooler months when temperatures are low and moisture is high because of overcast skies. The disease can continue well into the spring if these conditions persist.

Powdery mildew

Figure 124

Cause   *Oidium* sp.

Signs and symptoms   A white, powdery coating covers the tops and sometimes the bottoms of leaves on affected plants. The powdery coating sometimes forms in circular spots and sometimes covers the entire surface of the leaf.

Control   This disease can be serious in the interiorscape where the relatively dry conditions are ideal for powdery mildew disease. Triadimefon is very effective for powdery mildew control.

Rhizoctonia aerial blight

Figure 125

Cause   *Rhizoctonia solani*

Signs and symptoms   Rhizoctonia aerial blight occurs primarily during the summer months. Disease may develop in less than 1 week, so plants should be monitored carefully and frequently for initial symptoms. Brown, irregularly shaped spots form anywhere on the foliage but are most commonly in plant centers or near the soil where the inoculum originates. Sometimes the first lesions appear near the top of the plant, confusing the source of the pathogen. Lesions spread rapidly over the plants and cover them with the reddish brown, weblike mycelium of the pathogen.

Control   Cultural control of this disease is the same as that discussed for the other diseases. In addition, since the source of the pathogen can be the potting medium, the plants should be grown in new or disinfested pots and potting medium and on raised benches in an enclosed structure. Temperatures above 85°F (30°C) promote disease development, so cooling the greenhouse during certain times of year can aid disease control. PCNB and thiophanate methyl are moderately effective in *Rhizoctonia* control.

Codiaeum (Croton)

Crotons are colorful, woody shrubs native to the South Sea Islands and the Malay peninsula. They are used as foliage plants in dish gardens and terrariums and as floor plants. Plants are produced at 3,000 to 5,000 ft-c to develop optimum color contrast of leaves. Night temperatures should be maintained above 65°F (18°C) for good growth. Indoor light levels must be 150 ft-c or greater. Diseases of crotons include many caused by bacterial and fungal pathogens. Most common pests include mites, mealybugs, scales, and thrips.

Anthracnose

Figure 126

Cause   *Colletotrichum* sp. or *Glomerella cingulata*

Signs and symptoms   Spots form on croton leaves of all ages and are initially water soaked and become tan with time. Tiny, black, fungal fruiting bodies sometimes form in the dead tissue of the spot and may appear in concentric rings. Both the asexual (*Colletotrichum*) and sexual (*Glomerella*) stages are frequently isolated from leaf spots.

Control   Avoid wet foliage, since this is necessary for infection and spread of spores. This is especially crucial during mist propagation. Minimize plant handling because wounds allow spores of the pathogen to infect plants more easily. Increasing plant spacing, watering early in the day, and adding fans for increased air circulation reduce the relative humidity and consequently the conditions that are needed for spore production and dispersal. There are no fungicides labeled specifically for this use on crotons.

Selected Reference

Humphreys-Jones, D. R., and S. F. Flett. 1976. Leaf blotch (*Glomerella cingulata* (Stonem.) Spauls. & Schrenk) of *Codiaeum variegatum* (L.) Juss. Plant Pathol. 25:208.

Botrytis blight

Figure 127

Cause   *Botrytis cinerea*

Signs and symptoms   Large, tan to brown leaf spots with concentric rings are usually found at leaf edges and tips. Botrytis blight occurs primarily during cool periods of the year, especially on cuttings during shipping.

Control   Cultural controls are the same as those listed for anthracnose disease. The best fungicides for control of Botrytis blight are iprodione and vinclozolin, although they have been shown to occasionally cause stunting and slight distortion on some cultivars.

Crown gall

Figure 128

Cause   *Agrobacterium tumefaciens*

Signs and symptoms   Slightly swollen areas on the stems, leaf veins, and even roots are initial symptoms of disease. These swollen areas enlarge and become corky. Galls may also form on the ends of cuttings or on stems where cuttings have been removed. In cases of severe infection, galls may enlarge and merge to create a very distorted stem or root mass.

Control   Remove and destroy all plants found infected with the bacterium, and then disinfest cutting tools with alcohol, bleach, or quaternary ammonium. Since a fungus and some insects also cause galls on croton, an accurate laboratory diagnosis must be made in order to choose effective controls.

Fusarium root and stem rot

Figures 129 and 130

Cause   *Fusarium solani* and *F. oxysporum*

Signs and symptoms   Fusarium root and stem rot typically appears as a soft, mushy rot at the base of a cutting or rooted plant. The rotten area frequently has a purplish or reddish margin. *Fusarium solani* forms tiny, bright red, globular structures (fruiting bodies) at stem bases on severely infected plants. Roots are mushy and brown and easily disintegrate when handled. Sometimes leaves are infected during propagation.

Control If stem rot or cutting rot is a problem, treatment of the cuttings with a dip or a post-sticking drench of thiophanate methyl should diminish losses. Remove infected plants from stock areas as soon as they are detected. Since Fusarium root and stem rot is similar in appearance to many other root and stem rots, accurate disease diagnosis is very important prior to choice and application of fungicides.

Kutilakesa stem gall and canker

Figure 131

Cause *Kutilakesa pironii*

Signs and symptoms Stem galls on croton are large, subspherical, corky, roughened areas up to 5 cm found on stems in areas where cuttings have been removed. Galls are also found on petioles and leaf midveins. *Codiaeum* cultivar Elaine is highly resistant and cultivars Bravo, Norma, and Stoplight are highly susceptible to this pathogen. Wounded tissue appears to be a requirement for infection by *Kutilakesa pironii*.

Control Cultural controls are the same as those listed for crown gall. Predacious mycophagous mites are frequently associated with galled tissue and are perhaps responsible for spread of spores to new sites. Clean cutting instruments with a disinfestant such as bleach, alcohol, or quaternary ammonium to reduce spread during plant propagation.

Selected References

Alfieri, S. A., Jr., J. F. Knauss, and C. Wehlburg. 1979. A stem-gall and canker-inciting fungus, new to the United States. Plant Dis. Rep. 63:1016-1020.

Alfieri, S. A., Jr., C. L. Schoulties, and N. E. El-Gholl. 1980. *Nectriella (Kutilakesa) pironii*, a pathogen of ornamental plants. Proc. Fla. State Hortic. Soc. 93:218-219.

Stunting

Figure 132

Cause Overfertilization

Signs and symptoms The plant on the left in Figure 132 received an appropriate amount of fertilizer while that in the center received twice the recommended rate and the plant on the right received four times the recommended rate. In general, plants grow slowly when overfertilized and may have burned leaf tips or margins. This may be especially common when too low a light level is available or temperatures are low and plants cannot use the amount of fertilizer they are given.

Control Always use the recommended rate of fertilizer for the growing conditions. Trying to push this plant to grow faster results only in poor foliar color and slower growth. If the plants appear to be growing too slowly, be sure to check the concentration of soluble salts to determine whether they are under- or overfertilized. Check the root systems—poorly formed root systems with burned roots can be an indication of overfertilization.

Selected Reference

Chase, A. R., and R. T. Poole. 1989. Response of *Codiaeum variegatum* 'Gold Star' as influenced by slow-release fertilizer. J. Environ. Hort. 7:21-23.

Xanthomonas leaf spot

Figures 133 and 134

Cause *Xanthomonas campestris* pv. *poinsettiicola*

Signs and symptoms Foliar infections on croton start as tiny, pinpoint, water-soaked areas that can rapidly enlarge to 3 mm or more. They tend to remain confined to the areas between leaf veins and are very wet and dark brown or black appearing when well developed. On some cultivars, the spots have a bright yellow border. Most spots also show an irregularly shaped border, which is corky and especially visible on leaf undersides. All cultivars, as well as crown-of-thorns and poinsettia (also in the Euphorbiaceae family), are susceptible to this pathogen.

Control Eliminate all stock plants that have symptoms of Xanthomonas leaf spot. The disease is very difficult to control unless plants are produced without overhead watering or exposure to rainfall. Bactericides such as copper-containing compounds may be somewhat effective if used on a preventative and regular basis.

Selected Reference

Chase, A. R. 1985. Bacterial leaf spot of *Codiaeum variegatum* cultivars caused by *Xanthomonas campestris* pv. *poinsettiicola*. Plant Pathol. 34:446-448.

Coleus spp.

Coleus spp. are native to tropical Asia and Africa. They are extensively used as bedding plants as well as in small dish gardens. Coleus require 4,000 to 5,000 ft-c for good color and form and should be maintained with a minimum temperature of 60°F (15°C). Common diseases are damping-off, a few leaf spots, and a viral disease. Pests include caterpillars, mites, nematodes, scale insects, and white flies.

Botrytis blight

Figure 135

Cause *Botrytis cinerea*

Signs and symptoms Botrytis leaf and blossom blight usually appears on lower leaves of cuttings in contact with the potting medium. The water-soaked lesion may enlarge rapidly to encompass a large portion of the leaf blade or even the entire cutting. The area turns necrotic and dark brown to black with age. When night conditions are cool, day conditions warm, and moisture conditions high, the pathogen readily sporulates on both leaves and flowers, covering them with grayish brown, dusty masses of conidia.

Control Controlling Botrytis blight of foliage plants is particularly important during the winter months. Methods that improve foliage drying and reduce moisture condensation on foliage during the nights reduce the need for fungicide application. Chlorothalonil, iprodione, and vinclozolin are effective for *Botrytis* control.

Chimera

Figure 136

Cause Genetic variability in the plant

Signs and symptoms Leaves develop yellow variegation that can be mistaken for symptoms of viral infections. Plants with these symptoms occur rarely in a planting, unlike those infected with a virus, which may be in higher numbers throughout the crop. Distortion can also occur, especially on some cultivars produced in tissue culture where the plant genetics may become skewed because of the unusual growing conditions.

Control If the number of off-type plants is high, a new source of plants should be found. Discard those that are found, but don't miss the opportunity to develop a new selection of the plant. This is one of the oldest ways for new plants to come into the commercial trade.

Corynespora leaf spot

Figure 137

Cause *Corynespora cassiicola*

Signs and symptoms Lesions appear first as tiny, sunken, slightly brown areas, which enlarge to about 1 cm in diameter and darken with age. A bright purple or red margin and a chlorotic halo about 1 mm wide are usually present on this host. Leaf abscission is common under optimal conditions for disease expression. Similar symptoms are seen on other plants such as *Ficus, Nematanthus, Salvia, Columnea,* and *Saintpaulia ionantha.*

Control Use the same cultural controls as mentioned for Botrytis blight. Iprodione is very effective in disease control.

Pseudomonas leaf spot

Figure 138

Cause *Pseudomonas cichorii*

Signs and symptoms Spots are water soaked and turn dark green to black. They may have a yellow edge, but this is not common.

Control Avoid overhead watering as much as possible to reduce conditions for infection and spread of the pathogen.

Preventive applications of a copper-containing bactericide may reduce disease slightly, but none are labeled for coleus at this time. Many other hosts of *P. cichorii* have been identified: chrysanthemum, ferns, ficus, geranium, gerber daisy, and philodendron.

Root-knot nematode

Figure 139

Cause *Meloidogyne* spp.

Signs and symptoms Galls occur on roots, and the root system may be drastically reduced. Plant stunting and wilting occur when severe infestations are present.

Control Use sterile soil, and grow plants off the ground if possible. Check nematicide labels for this plant and application methods.

Sunscald

Figure 140

Cause High light levels

Signs and symptoms Leaves rapidly develop tan to white, irregularly shaped areas. The spots appear overnight after plants are exposed to light brighter than that to which they are acclimated.

Control Never move a plant from low-light or shaded conditions into very bright or direct sunlight. If plants must be moved to a location with higher light, make the move in increments from low to medium to high light over a period of weeks to allow the leaves to adjust to the higher light levels.

Columnea

Columnea are used primarily in hanging baskets in the interiorscape. Plants are produced under 1,500 to 3,000 ft-c with minimum night temperatures of 60°F (15°C). Soil temperatures below 65°F (18°C) will cause growth to slow down or stop. Diseases of *Columnea* are caused primarily by fungi that attack roots, stems, and foliage. Spider and tarsonemid mites, mealybugs, and scales are common pests.

Corynespora leaf spot

Figure 141

Cause *Corynespora cassiicola*

Signs and symptoms Lesions appear first as tiny, sunken, slightly brown areas, which enlarge to about 1 cm in diameter and darken with age. A bright purple or red margin and a chlorotic halo about 1 mm wide are sometimes present. Similar symptoms are seen on other gesneriads such as *Aeschynanthus* spp., *Aphelandra squarrosa, Nematanthus,* and *Saintpaulia ionantha* (African violet), as well as some species of *Ficus.*

Control Eliminate overhead irrigation and exposure to rainfall, and use only pathogen-free stock for cuttings. Chemical control trials have indicated that both mancozeb and chlorothalonil provide excellent disease control on lipstick vine.

Selected Reference

Chase, A. R. 1982. Corynespora leaf spot of *Aeschynanthus pulcher* and related plants. Plant Dis. 66:739-740.

Rhizoctonia aerial blight

Figure 142

Cause *Rhizoctonia solani*

Signs and symptoms Leaves turn brown and become matted together. Rooting cuttings may be completely covered by a mass of brownish mycelia. Growth of mycelia from the potting medium onto larger plants can escape notice and give the appearance that plants have been infected from an aerial source of inoculum. Close examination, however, generally reveals the presence of mycelia on stems prior to development of obvious foliar symptoms. In severe infections, affected leaves drop from the stems, giving the plant a barren appearance. This disease is most common during the hottest times of the year when the plant foliage remains wet for long periods or relative humidity is high.

Control Use pathogen-free cuttings and new pots and potting media, and avoid extremes in soil moisture. Chemical control of diseases caused by *Rhizoctonia* has been investigated on many plants with a variety of fungicides. The fungicide used most widely as a soil drench to control *Rhizoctonia* diseases is thiophanate methyl.

Cordyline (Ti plants)

Ti plants are native to tropical and subtropical regions of Australia and Asia. One species of *Cordyline* is used in the Hawaiian islands as a wrapper for native foods. Although ti plants grow to the size of trees in their native environment, they are generally used as large, potted, floor plants or table-top plants indoors. Light intensities of 3,000 to 3,500 ft-c are used for production of ti plants, with a minimal night temperature of 65°F (18°C). Light levels between 100 and 150 ft-c are sufficient for use indoors. Many fungal and bacterial diseases occur on these plants during propagation and production. Mites are perhaps the most common pest of ti plants.

Anthracnose

Figure 143

Cause *Colletotrichum* sp. or *Glomerella cingulata*

Signs and symptoms Spots form on leaves of all ages and are initially water soaked and become tan with age.

Tiny, black, fungal fruiting bodies sometimes form in the dead tissue of the spot and may appear in concentric rings. Isolations from infected leaves may yield either *Colletotrichum* (asexual stage) or *Glomerella* (sexual stage).

Control Avoid wet foliage, since this is necessary for infection and spread of spores. This is especially crucial during mist propagation. Use only pathogen-free sources for cuttings. Water early in the day, increase plant spacing, and use fans to allow leaves to dry as quickly as possible. Mancozeb is effective in controlling anthracnose on ti plants.

Cercospora leaf spot

Figure 144

Cause *Cercospora* sp.

Signs and symptoms Initially rust-colored specks form. Many specks merge to form rectangular areas between the leaf veins. All tested varieties of ti plants are susceptible to *Cercospora* sp. Infection occurs through the stomata, and symptoms appear approximately 15 to 25 days later.

Control Use the same controls as those listed for anthracnose.

Selected Reference

Buddenhagen, I., and A. H. McCain. 1967. False rust of Ti (*Cordyline terminalis*), a new disease caused by an undescribed *Cercospora* species. (Abstr.) Phytopathology 57:97.

Erwinia blight

Figures 145, 146, and 147

Cause *Erwinia chrysanthemi, E. carotovora* pv. *carotovora*

Signs and symptoms Spots on leaves and stems are usually water soaked and slimy; eventually the spots coalesce and tissue disintegration occurs. Severe infections can result in cutting loss, since the plants often rot from the cutting end upward.

Control Carefully examine all cuttings used for propagation, destroying those that are suspected of having Erwinia blight. Recutting diseased propagative material to remove diseased portions only postpones the loss of that material from root rot and increases the chances of spread to healthy material. Always use pathogen-free plants for stock as well. There are no bactericides that provide an appreciable degree of control of *Erwinia* diseases.

Fluoride toxicity

Figure 148

Cause Excess fluoride

Signs and symptoms The first indication of fluoride toxicity on *Cordyline* is tipburn, which is followed by marginal necrosis. In severe cases, mottling also occurs within the center of the leaf and the entire leaf may die. *Cordyline terminalis* 'Baby Doll' is the cultivar most susceptible to this problem, although all cultivars have been observed to have symptoms when fluoride is present in water, soil, or fertilizer.

Control Where fluoride is known to be a problem, the propagation and potting media should have a pH of 6.0 to 6.5 to reduce fluoride availability. Cuttings without roots are likely to take up large amounts of fluoride. Some producers utilize treated water in propagation areas to prevent fluoride uptake during rooting and untreated water after rooting. To reduce fluoride uptake, use potting media, irrigation water, and fertilizers low in fluoride content. Irrigation water should contain less than 0.25 ppm of fluoride.

Selected References

Conover, C. A., and R. T. Poole. 1971. Influence of fluoride on foliar necrosis of *Cordyline terminalis* cv. Baby Doll during propagation. Proc. Fla. State Hortic. Soc. 84:380-383.
Poole, R. T., and C. A. Conover. 1973. Fluoride induced necrosis of *Cordyline terminalis* Kunth 'Baby Doll' as influenced by medium and pH. J. Am. Soc. Hortic. Sci. 98:447-448.

Fusarium stem and root rot

Figures 149 and 150

Cause *Fusarium* spp.

Signs and symptoms Fusarium root and stem rot typically appears as a soft, mushy rot at the base of a cutting or rooted plant. The rotten area frequently has a purplish or reddish margin. *Fusarium solani* forms tiny, bright red, globular structures (fruiting bodies) at stem bases on severely infected plants. Roots are mushy and brown and easily disintegrate when handled.

Control If stem rot or cutting rot is a problem, treatment of the cuttings with a dip or a post-sticking drench of thiophanate methyl should diminish losses. Remove infected plants from stock areas as soon as they are detected. Since Fusarium root and stem rot is similar in appearance to many other root and stem rots, accurate disease diagnosis is very important prior to choice and application of fungicides.

Phyllosticta leaf spot

Figure 151

Cause *Phyllosticta dracaenae*

Signs and symptoms Spots are circular to slightly irregular and range from pinpoint to slightly larger than 1 cm in diameter. They appear mainly on the older leaves of plants and are usually tan with purple borders and yellow halos. Under conditions of high disease pressure, the spots may run together and the entire leaf may die.

Control Use the same controls as those mentioned for anthracnose.

Selected References

Desai, M. V., and K. P. Patel. 1961. Leaf blight of *Dracaena* incited by *Phyllosticta draconis*. Plant Dis. Rep. 45:203.
Seymour, C. P. 1974. Phyllosticta leaf spot of *Dracaena*. Fla. Dept. Agric. & Cons. Serv. Plant Pathol. Circ. No. 143.

Southern blight

Figure 152

Cause *Sclerotium rolfsii*

Signs and symptoms The pathogen generally attacks the crown of the plant first, sometimes leaving the roots intact and causing girdling and collapse of the tops only. Sclerotia, the survival bodies of the pathogen, form all over the infected tissue and appear as small, mustard seed-sized bodies, which are initially white and turn brown as they mature. The white, fanlike mycelium of the pathogen also forms over the plant, potting medium, and even sides of benches.

Control Grow plants on raised benches and check cuttings carefully for signs of southern blight before sticking. Always discard plants suspected of southern blight infection, and use pathogen-free potting media and pots because the organism lives in soil and can be transferred from one crop to the next on recycled materials and equipment. Use of PCNB has been effective in controlling southern blight on some plants. Do not use PCNB as a preventative, since this fungicide can occasionally cause stunting when used repeatedly on some plants.

Xanthomonas leaf spot

Figure 153

Cause *Xanthomonas campestris*

Signs and symptoms Symptoms are generally confined to pinpoint yellow to tan lesions scattered across the leaf surface, although they can become large and confined between leaf veins. Spots are mostly 1 to 2 mm wide with irregularly raised edges. Severe infections can cause distortion of new leaves as well as complete yellowing and collapse of older leaves.

Control Eliminate all stock plants that have Xanthomonas leaf spot. The disease is very difficult to control unless plants are produced without overhead watering or exposure to rainfall. Bactericides such as copper-containing compounds will rarely be effective, even if used on a preventative and regular basis.

Dieffenbachia

Dieffenbachias are herbaceous plants native to Central and South America. Plants in production range from 15 cm to more than 1 m tall. Interior use is limited for many cultivars because of a relatively high light requirement: a few cultivars or species do well at 100 ft-c for 12 hr/day. Dieffenbachias are grown with 1,500 to 2,500 ft-c and up to 4,000 ft-c for production of cane as propagation stock (cane production has diminished for many species because of the advent of tissue-cultured plants). Many cultivars of dieffenbachia are currently propagated through tissue-culture techniques. Temperatures below 65 to 70°F (18 to 21°C) result in poor growth and rooting of cuttings. Dieffenbachias suffer from many diseases caused by viruses, bacteria, and fungi. They are commonly infested with mites, scales, mealybugs, and aphids; some of these pests affect the roots, and others affect leaves and stems.

Anthracnose

Figure 154

Cause *Colletotrichum gloeosporioides*, *Glomerella cingulata*

Signs and symptoms Symptoms occur primarily during the cooler, winter months. Leaf spots are initially tan and water soaked and may have a bright yellow halo. Fruiting bodies of the pathogen appear in concentric rings of tiny black specks within the leaf spot. As with some other anthracnose diseases, the sexual stage (*Glomerella*) or the asexual stage (*Colletotrichum*) of the pathogen may be isolated from infected leaves.

Control Keep foliage dry, and protect from cold water drips caused by condensation on overhead structures. Scout plants regularly, and remove those (or at least the individual leaves) with symptoms of anthracnose. Many fungicides, such as thiophanate methyl, mancozeb, and iprodione, provide some control of anthracnose on dieffenbachia.

Selected Reference

Semer, C. R., IV, B. C. Raju, and B. L. Tepper. 1983. A new disease of *Dieffenbachia maculata* (Lodd.) G. Don 'Camille': Leaf spot caused by *Colletotrichum gloeosporioides* (Penz.) Sacc. during transit and its control by the application of fungicides. Proc. Fla. State Hortic. Soc. 96:280-282.

Cold temperature damage

Figure 155

Cause Cold air during shipment

Signs and symptoms Areas between the main veins become chlorotic or light brown as a result of low temperatures.

Control Prevent exposure to low temperatures. Avoid extreme or abrupt changes in temperature. Although some dieffenbachias can be exposed to 45°F (7°C) without apparent leaf damage, plants grown at high temperatures (70 to 95°F [21 to 35°C]) can be damaged if the temperature drops to 50°F (13°C) or below.

Copper toxicity

Figure 156

Cause Application of copper compounds in an acid solution

Signs and symptoms Leaves develop tiny, water-soaked areas scattered across their surfaces. Spots may be irregularly shaped and mimic those caused by bacteria. Spots become corky in appearance and tan to brown within 2 to 3 weeks of application of a copper product.

Control Copper compounds are safely used on dieffenbachias when applied at recommended rates and intervals. Only when a copper compound is mixed with an acidic compound such as vinegar or the fungicide fosetyl aluminum will copper cause toxicity. Avoid applications of copper and acidic compounds to the same plant unless the interval exceeds 1 week. If the water pH is in the acidic range, adjust it to a pH of 6 before using it to apply compounds with copper.

Selected Reference

Chase, A. R. 1989. Aliette 80WP and bacterial disease control. Phytotoxicity. Foliage Digest 12(11):4-5.

Dasheen mosaic

Figures 157, 158, and 159

Cause Dasheen mosaic virus (DMV)

Signs and symptoms Dasheen mosaic virus is most severe on the cultivar Perfection and related cultivars of dieffenbachia. Symptoms that include ring spots, mosaic, leaf distortion, and stunting appear periodically during the year.

Control DMV is spread by both aphids and human handling, although the latter is the more common vector. It is very important to use pathogen-free stock, since the symptoms of DMV are not always noticeable. Table 10 lists

Table 10. Susceptibility of some dieffenbachias to *Xanthomonas campestris* pv. *dieffenbachiae* and Dasheen mosaic virus

Species or cultivar	*Xanthomonas*	Dasheen mosaic virus
amoena	Not tested	Slight
× *bausei*	Slight	Hypersensitive, dies
Camille	Slight	Chronic, severe
Compacta	Slight	Chronic, severe
maculata	Not tested	Moderate
× *memoria-corsii*	Slight	Hypersensitive, dies
Perfection	Slight	Chronic, severe
Rudolph Roehrs	Not tested	Moderate
Star White	Slight	Not tested
Starry Nights	Slight	Not tested
Triumph	Moderate	Not tested
Tropic Star	Resistant	Not tested
Victory	Moderate	Not tested

responses of some *Dieffenbachia* species and cultivars to DMV infection. Currently available chemicals do not affect viruses. Other hosts such as *Aglaonema*, *Philodendron*, and *Spathiphyllum* must be monitored for symptoms, since they can act as reservoirs for the virus.

Selected Reference

Chase, A. R., and F. W. Zettler. 1982. Dasheen mosaic virus infection of *Dieffenbachia* cultivars. Plant Dis. 66:891-893.

Erwinia blight and stem rot

Figures 160 and 161

Cause   *Erwinia carotovora* pv. *carotovora* and *E. chrysanthemi*

Signs and symptoms   Stem rot caused by *Erwinia* spp. is very similar in appearance to that caused by *Fusarium* and *Phytophthora* spp. Rotted areas are usually watery and mushy and have a rotten, fishy odor in many cases. The bacteria sometimes form a slimy, gelatinous mass at the base of infected cuttings, and infected plants generally have yellow lower leaves. Leaf spots caused by *Erwinia* spp. enlarge rapidly, and centers may become so watery that they fall out.

Control   The only successful means of controlling this disease is the eradication of symptomatic plants. This should be done during the hot months when Erwinia blight is most likely to appear. Use of infected plants that are not showing symptoms (asymptomatic) will generally result in cutting loss, since the bacterium is found inside the plant stem (systemic) and becomes active during rooting. Antibiotic and copper compounds may provide limited control of the leaf spot symptom. Keep plant foliage dry to minimize new infections. Most other foliage plants are susceptible to this bacterium and must be considered potential sources of infection for dieffenbachia.

Selected References

Dickey, R. S. 1981. *Erwinia chrysanthemi*: Reaction of eight plant species to strains from several hosts and to strains of other *Erwinia* species. Phytopathology 71:23-29.

McFadden, L. A. 1961. Bacterial stem and leaf rot of *Dieffenbachia* in Florida. Phytopathology 51:663-668.

Munnecke, D. E. 1960. Bacterial stem rot of *Dieffenbachia*. Phytopathology 50:696-700.

Nieves-Brun, C. 1985. Infection of roots of *Dieffenbachia maculata* by the foliar blight and soft rot pathogen, *Erwinia chrysanthemi*. Plant Pathol. 34:139-145.

Fusarium leaf spot and stem rot

Figures 162 and 163

Cause   *Fusarium solani*

Signs and symptoms   Fusarium stem rot typically appears as a soft, mushy rot at the base of a cutting or rooted plant. The rotten area frequently has a purplish or reddish margin. Infection of leaves under very wet conditions results in tan, papery leaf spots with concentric rings of light and dark tissue. *Fusarium solani* forms tiny, bright red, globular structures (perithecia are the sexual stage [fruiting bodies] of this pathogen) at stem bases on severely infected plants.

Control   If stem rot or cutting rot is a problem, treatment of the cuttings with a dip or a post-sticking drench of thiophanate methyl should diminish losses. Remove infected plants from stock areas as soon as they are detected. Since Fusarium stem rot is similar in appearance to Erwinia stem rot, accurate disease diagnosis is very important prior to choice and application of pesticides.

Selected Reference

Chase, A. R., and N. E. El-Gholl. 1982. Stem rot, cutting rot, and leaf spot of *Dieffenbachia maculata* 'Perfection' incited by *Fusarium solani*. Plant Dis. 66:595-598.

Leptosphaeria or brown leaf spot

Figure 164

Cause   *Leptosphaeria* sp.

Signs and symptoms   Lesions appear initially as tiny, water-soaked areas scattered across the leaf surface. These lesions may be very numerous (50 per leaf) and are usually less than 1 mm wide. Lesions are not delimited by the veins and occasionally occur on the midvein, petioles, and flower spathes. Mature lesions are roughly circular and tan and have an orange yellow halo. The ascocarps of the pathogen frequently form in the lesions and appear as tiny, black specks. Spots on some cultivars are surrounded by a water-soaked margin. Affected leaves become chlorotic and droop but do not usually abscise.

Control   The following species and cultivars of *Dieffenbachia* were found equally susceptible to *Leptosphaeria* sp. in greenhouse inoculation trials: *D. amoena*, *D. maculata*, *D. maculata* 'Exotica,' *D. maculata* 'Rudolph Roehrs,' and *D. × memoria-corsii*. No infection occurred on any of these hosts without wounding. Mature leaves are significantly more susceptible to this pathogen than immature leaves, and lower leaf surfaces are more susceptible than upper surfaces. Continuous relative humidity of 100% appears to reduce disease severity compared with variable relative humidities between 78 and 92%. Chemical controls have not been investigated for this disease.

Selected Reference

Marlatt, R. B. 1966. Brown leaf spot of *Dieffenbachia*. Plant Dis. Rep. 50:687-689.

Myrothecium leaf spot and petiole rot

Figure 165

Cause   *Myrothecium roridum*

Signs and symptoms Myrothecium leaf spot most frequently appears on wounded areas of leaves, such as tips and breaks in the main vein that occur during handling. The leaf spots are watery and nearly always contain the black and white fungal fruiting bodies in concentric rings near the outer edges on the leaf undersides. The presence of these bodies is good evidence that the cause is *Myrothecium*. Newly planted, tissue-cultured explants are especially susceptible to this disease. The primary symptom on these explants is petiole rot starting with the oldest leaves, although leaf spot can occur as well.

Control Avoid wounding leaves, and keep the foliage as dry as possible. Many other plants are hosts of *M. roridum*, including *Aglaonema, Aphelandra, Begonia, Calathea, Spathiphyllum,* and *Syngonium,* and these plants must be included in control programs. *Myrothecium* diseases are most severe at temperatures between 70 and 81°F (21 and 27°C) and on overfertilized dieffenbachias. Preventive fungicide treatments to newly transplanted, tissue-cultured plants may be necessary when disease pressure is high. Iprodione provides some control of the leaf spot phase of this disease and is very effective in tissue-cultured plantlets affected with petiole rot. Chlorothalonil also gives good control of Myrothecium leaf spot but is not very effective for control of petiole rot. In addition, mancozeb provides good control of the leaf spot on some hosts.

Selected References

Chase, A. R. 1983. Influence of host plant and isolate source on Myrothecium leaf spot of foliage plants. Plant Dis. 67:668-671.

Chase, A. R., and R. T. Poole. 1984. Development of Myrothecium leaf spot of *Dieffenbachia maculata* 'Perfection' at various temperatures. Plant Dis. 68:488-490.

Chase, A. R., and R. T. Poole. 1985. Host nutrition and severity of Myrothecium leaf spot of *Dieffenbachia maculata* 'Perfection'. Scientia Horticulturae 25:85-92.

Phytophthora stem rot and leaf spot

Figure 166

Cause *Phytophthora parasitica, P. palmivora*

Signs and symptoms This disease occurs primarily on plants grown in or on the ground in south Florida. Leaf spots are initially small and water soaked with irregular margins. They may become tan and papery if conditions are dry, or their centers may fall out if conditions are wet. Stem rot usually begins at the soil line where the stem becomes soft and watery and lower leaves turn yellow. Eventually, the area becomes sunken and a cavity may form, resulting in lodging of the stem.

Control Growing plants on raised benches, away from the natural source of infection (the soil), is the best way to avoid this disease. Use new or sterilized pots and potting medium. Poorly draining potting media can increase severity of this disease. Because of the similarities between this and several other diseases, diagnosis must be confirmed by a diagnostic laboratory before optimum control strategies can be chosen. A combination of thiophanate methyl and etridiazole, etridiazole alone, metalaxyl, and fosetyl aluminum provide the best disease control of stem rot when cultural control methods are also employed. Fosetyl aluminum also can be applied as a foliar spray for leaf spot control.

Selected References

Ridings, W. H., and J. J. McRitchie. 1974. Phytophthora leaf spot of *Philodendron oxycardium* and related species. Proc. Fla. State Hortic. Soc. 87:442-447.

Tompkins, C. M., and C. M. Tucker. 1947. Stem rot of *Dieffenbachia picta* caused by *Phytophthora palmivora* and its control. Phytopathology 37:868-874.

Pythium root rot

Figure 167

Cause *Pythium* spp. (mainly *P. splendens*)

Signs and symptoms Root rot caused by *Pythium* sp. is usually first noticed when the lower leaves of cuttings or plants turn yellow and wilt. Examination of the roots reveals that they are mushy and brown and usually sparse.

Control Use the same controls as those mentioned for Phytophthora stem and root rot.

Selected Reference

Tisdale, W. B., and G. D. Ruehle. 1949. Pythium root rot of aroids and Easter lilies. Phytopathology 39:167-170.

Sunburn

Figure 168

Cause High light levels

Signs and symptoms Leaves rapidly develop tan to white, irregularly shaped areas. The spots appear overnight after the plants are exposed to light much brighter than that to which they are acclimated.

Control Never move a plant from a low-light or shaded condition into very bright or direct sunlight. If plants must be moved to higher light, make the move in increments from low to medium to high light over a period of weeks to allow the leaves to adjust to the higher light levels.

Xanthomonas leaf spot

Figure 169

Cause *Xanthomonas campestris* pv. *dieffenbachiae*

Signs and symptoms Foliar infections on dieffenbachia start as tiny, pinpoint, water-soaked areas that can rapidly enlarge to 3 mm or more. They tend to form on leaf margins where the bacterium can enter the leaf through hydathodes. When a main vein in the leaf is invaded, the infection rapidly spreads throughout the leaf. These necrotic areas are frequently very black and surrounded by a bright yellow halo. Some cultivars are more resistant than others (Table 10). Most other plants in the Aroid family, such as *Aglaonema, Anthurium,* and *Syngonium,* are also subject to this disease.

Control   Eliminate all stock plants that have Xanthomonas leaf spot. The disease is very difficult to control unless plants are produced without overhead watering or exposure to rainfall. Xanthomonas blight of dieffenbachia is most severe at temperatures between 86 and 91°F (30 and 35°C). Bactericides such as copper-containing compounds may be somewhat effective if used on a preventative and regular basis. Nutritional studies on dieffenbachias have shown that applications of fertilizer at rates greater than recommended result in decreased susceptibility to *X. campestris* pv. *dieffenbachiae* without causing appreciable damage to the appearance of the plant.

Selected References

Chase, A. R. 1990. High fertilizer rates reduce Xanthomonas leaf spot of *Dieffenbachia.* Foliage Digest 13(8):3-5.

Chase, A. R., and R. J. Henny. 1989. Sensitivity of 12 *Dieffenbachia* cultivars to *Xanthomonas.* Foliage Digest 13(2):1-2.

McCulloch, L., and P. P. Pirone. 1939. Bacterial leaf spot of *Dieffenbachia.* Phytopathology 29:956-962.

Dionaea (Venus's fly-trap)

These insectivorous plants are used primarily in small dish gardens or terrariums. They are very sensitive to temperatures below 41°F (5°C) and must be kept constantly moist. They are susceptible to a variety of fungal diseases when produced commercially.

Anthracnose

Figure 170

Cause   *Colletotrichum gloeosporioides*

Signs and symptoms   Water-soaked, longitudinal streaks with yellow borders are the first symptoms. As the lesions age, they turn gray with a darker center and eventually become elliptical, black spots on the petioles and leaf blades. They commonly merge to encompass large portions of the plant.

Control   Keep foliage as dry as possible. Mancozeb and chlorothalonil were effective in controlling this disease experimentally but may not be labeled for use on Venus's fly-trap.

Selected Reference

Knauss, J. F. 1972. Foliar blight of *Dionaea muscipula* incited by *Colletotrichum gloeosporioides.* Plant Dis. Rep. 56:391-393.

Pythium root rot

Figure 171

Cause   *Pythium* spp.

Signs and symptoms   Leaves wilt, may turn yellow or pale green, and eventually die. Plants are frequently stunted, and examination of roots reveals their rotted condition. Initial infections of the roots appear as small, water-soaked, grayish or brown areas. These spots can rapidly expand to affect the entire root system. Severely infected plants may have no living roots remaining by the time they are examined.

Control   Prevention is always the best control of a soil-borne pathogen like *Pythium.* Use clean pots and potting media, and grow plants on raised benches. Fungicides that are effective in controlling most Pythium root rot diseases include etridiazole, metalaxyl, and fosetyl aluminum. Check labels for legal use on Venus's fly-trap.

Dracaena

Dracaenas are the most commonly produced member of the Agavaceae family. Most are native to the tropical regions of Africa and Asia and vary in size from less than 0.3 m to trees exceeding 3 m. Use as foliage plants includes everything from additions to small dish gardens to large specimen plantings of corn plant (*D. massangeana*) reaching 3 m. Dracaenas are grown under a variety of light levels ranging from full sun (*D. marginata*) to 63% shade (*D. fragrans*) and 73% shade (*D. deremensis*). Interiorscape light levels of 75 to 150 ft-c are recommended. The most common diseases of dracaenas are caused by fungi. Mites can be severe on small plants, and scales occur on several cultivars of dracaenas.

Anthracnose

Figure 172

Cause   *Colletotrichum* spp.

Signs and symptoms   Anthracnose is characterized by yellow and later dark brown spots anywhere on the leaf. Yellowish masses of spores form in zones along leaf veins or in concentric rings in the spot. Eventually leaves may collapse entirely and die. *Dracaena sanderana* cultivars are commonly infected with this pathogen during the summer months and appear especially susceptible when they are being rooted under mist conditions.

Control   Keep plant stresses from water and heat to a minimum. Do not use any cuttings that have spots. On rooted plants, minimize overhead irrigation and exposure to rainfall. Fungicides that are effective for controlling anthracnose include mancozeb, iprodione, and chlorothalonil.

Selected Reference

Graham, S. O., and J. W. Strobel. 1958. The incidence of anthracnose fungi on ornamental foliage plants in Washington State greenhouses. Plant Dis. Rep. 42:1294-1296.

Botrytis blight

Figure 173

Cause *Botrytis cinerea*

Signs and symptoms Spots usually appear on the leaf underside, especially on petioles near the pot rim or in contact with the potting medium. A small, water-soaked spot can rapidly enlarge and cover the entire leaf. Sporulation on necrotic leaves or flowers appears as a powdery, grayish brown mass.

Control Watch for *Botrytis* when the following conditions occur: low light, high humidity, poor air circulation, and warm days with cool nights. Improve air circulation and water early in the day to facilitate rapid drying of foliage. Venting and heating the greenhouse at sundown can also reduce humidity overnight and thus reduce development of Botrytis blight. Fungicides containing vinclozolin or iprodione effectively control Botrytis blight.

Cercospora leaf spot

Figure 174

Cause *Cercospora* sp.

Signs and symptoms Tiny, slightly raised, red or dark green spots appear on lower leaf surfaces. Spots enlarge slowly and eventually are visible on the tops of leaves as indiscreet yellow and brown areas.

Control Use the cultural controls listed for anthracnose. Thiophanate methyl or a combination of thiophanate methyl and mancozeb is effective.

Cold damage

Figure 175

Cause Cold air during shipment

Symptoms Areas between main veins or along leaf margins become chlorotic or light brown as a result of low temperatures.

Control Prevent exposure to low temperatures. Avoid extreme or abrupt changes in temperature. Although some dracaenas can be exposed to 45°F (7°C) without apparent leaf damage, plants grown at high temperatures (70 to 95°F [21 to 35°C]) can be damaged if the temperature drops to 45°F (7°C) or below.

Selected References

Hodyss, L. B. 1988. Analysis of shipping related problems of *Dracaena massangeana*. Proc. Fla. State Hortic. Soc. 101:89-90.

Marousky, F. J. 1980. Chilling injury in *Dracaena sanderana* and *Spathiphyllum* 'Clevelandii'. HortScience 15:197-198.

Poole, R. T., C. A. Conover, and P. G. Webb. 1987. Effect of environmental factors on *Dracaena* 'Massangeana' during shipping. Proc. Fla. State Hortic. Soc. 100:340-341.

Erwinia blight

Figures 176, 177, and 178

Cause *Erwinia carotovora* pv. *carotovora* and *E. chrysanthemi*; rarely *E. herbicola*

Signs and symptoms Stem ends of unrooted and sometimes rooted cuttings are mushy and brown and frequently smell like rotted fish. The ends are wet and disintegrate rapidly under the warm, moist conditions of rooting beds. A bacterial slime is sometimes present. Infection commonly extends into the lower leaves and causes their discoloration and collapse. If the ends of infected stems are cut, a darkened vascular system may be seen. Discrete spots that are initially water soaked, dark, and rounded form when plants are infected with *E. herbicola*. These spots rarely enlarge to more than 3 mm across, while those caused by the other species of *Erwinia* can continue to enlarge until the entire leaf is affected.

Control The practice of recutting infected plants to remove rotted portions does little to diminish losses of cuttings. Sometimes cuttings will root only to become symptomatic and rotted at a later date. Dipping asymptomatic cuttings sometimes increases disease losses, even when copper or streptomycin products are employed, because the bacteria are easily spread in dip solutions. The only way to eliminate this disease is to reject all cuttings with these symptoms.

Selected Reference

Chase, A. R. 1984. Leaf spot diseases of *Dracaena sanderana* caused by two species of *Erwinia*. Plant Dis. 68:251.

Flecking

Figure 179

Cause Unknown

Signs and symptoms The newest leaves have scattered, white to yellow spots that are most common near the apex in *D. marginata*. As the leaf matures, these spots usually turn green, but marketability is reduced when plants are severely spotted. The disease is most prevalent during the winter, most severe under high light levels (minimal at 80% shade), and does not appear to be affected by fertilizer level or irrigation frequency.

Control The problem is most severe on plants grown in high light and low temperatures and decreases as the percentage of shade is increased. Plants grown at 2,000 ft-c or less often have no spotting, while plants grown with full sun often have spotting.

Selected Reference

Poole, R. T., and C. A. Conover. 1985. Flecking of *Dracaena marginata*. Nurserymen's Digest 19(6):57.

Fluoride toxicity

Figures 180 and 181

Cause   Excess fluoride

Signs and symptoms   On Massangeana, the first indication is a mottled loss of pigmentation within the green area, most easily observed on the undersides of leaves. With time, these areas become chlorotic and then necrotic, often progressing to the point that the leaf margin is damaged. Elongated, tan to dark brown areas form first in the white tissue of the cultivar Warneckii and then progress to marginal necrosis. Tips and margins of leaves of the cultivar Janet Craig have chlorotic and necrotic areas.

Control   Potting media should have a pH of 6.0 to 6.5 to control fluoride availability. Care must also be taken to use water, potting media, and fertilizer sources that do not contain fluoride. Superphosphate and perlite have been shown to contain significant levels of fluoride and should be avoided when growing fluoride-sensitive plants. Dracaenas are also very sensitive to fluoride air pollution.

Selected References

Conover, C. A., and R. T. Poole. 1980. Influence of fertilization, superphosphate and lime on mottling of *Dracaena fragrans*. HortScience 15:23-24.

Conover, C. A., and R. T. Poole. 1982. Fluoride induced chlorosis and necrosis of *Dracaena fragrans* 'Massangeana'. J. Am. Soc. Hortic. Sci. 107:136-139.

Poole, R. T., and C. A. Conover. 1975. Fluoride-induced necrosis of *Dracaena deremensis* Engler cv. Janet Craig. HortScience 10:376-377.

Poole, R. T., and C. A. Conover. 1974. Foliar chlorosis of *Dracaena deremensis* Engler cv. Warneckii cuttings induced by fluoride. HortScience 9:378-379.

Fusarium leaf spot and stem rot

Figures 182, 183, and 184

Cause   *Fusarium moniliforme*

Signs and symptoms   Fusarium leaf spot occurs initially on the newest leaves in the central whorl when it is wet. Lesions are irregularly shaped and tan to reddish brown and many times have a chlorotic (yellow) border. Under conditions of high disease pressure and continually wet foliage, the lesions coalesce and infection spreads into the plant meristem. Stem rot often occurs on cuttings during mist propagation. Symptoms are identical to those caused by *Erwinia* spp.; culture of the pathogen is required to differentiate the two diseases. The creamy orange spores of the pathogen are commonly produced in leaf or stem lesions under wet conditions, and splashing water spreads them onto other plants.

Control   Keeping the foliage of this plant dry can eliminate the foliar phase of this disease. If this is not possible, use one of several fungicides to diminish symptom expression. Chlorothalonil and mancozeb provide excellent control of Fusarium leaf spot of dracaenas. Disease is most severe when temperatures are between 60 and 80°F (15 and 27°C). Table 11 gives the susceptibility of many species and cultivars of dracaena and related plants to Fusarium leaf spot. In addition, red-edge dracaenas that receive excess fertilizer are somewhat resistant to this disease.

Selected References

Chase, A. R. 1987. Effect of fertilizer level on severity of Fusarium leaf spot of *Dracaena marginata*. Proc. Fla. State Hortic. Soc. 100:360-362.

Chase, A. R. 1993. Fusarium leaf spot of dracaenas—Resistance of species and cultivars. CFREC-Apopka Research Report, RH-93-10.

TABLE 11. Resistance of selected *Dracaena* species and cultivars to *Fusarium moniliforme*, the cause of Fusarium leaf spot

Dracaena species Cultivar	Resistant or susceptible	Symptom type
deremensis		
'Compacta'	Resistant	None
'Janet Craig'	Very slightly susceptible	Clear, yellow speckles on leaf edges
'Warneckii'	Very slightly susceptible	Clear, yellow speckles on leaf edges
'Lemon Lime'	Resistant	None
fragrans 'Massangeana'	Very slightly susceptible	Small, yellow spots on leaf edges
marginata	Very highly susceptible	Large, yellow and brown spots in the whorl that merge to rot the center
'Bicolor'	Very slightly susceptible	Small, yellow spots in the whorl
'Colorama'	Moderately susceptible	Small, yellow spots in the whorl
'Magenta'	Very highly susceptible	Large, brown spots in the whorl that merge to rot the center
'Tricolor'	Slightly susceptible	Small, yellow spots in the whorl
reflexa 'Song of Jamaica'	Highly susceptible	Large, tan to reddish spots in the whorl and on leaf edges
sanderana	Resistant	None
'Borinquensis'	Resistant	None
'Gold'	Resistant	None
surculosa	Very slightly susceptible	Large, tan, papery spots form where the leaf joins the stem
'Florida Beauty'	Very slightly susceptible	Large, tan, papery spots form where the leaf joins the stem
'Juanita'	Very slightly susceptible	Large, tan, papery spots form where the leaf joins the stem

Knauss, J. F. 1971. Fusarium stem rot, a previously unreported disease of unrooted cuttings of *Dracaena*. Proc. Am. Soc. Hortic. Sci., Trop. Reg. 15:208-215.

Wehlburg, C., and A. P. Martinez. 1967. Leaf spot of *Dracaena marginata* Lam. caused by *Fusarium moniliforme* Sheld. and its control. Proc. Fla. State Hortic. Soc. 80:454-456.

Notching

Figure 185

Cause   High temperatures

Signs and symptoms   The bases of leaves on *Dracaena* 'Warneckii' appear to be cut by a knife. The serrations are perpendicular to the long axis of the leaf and 1 to 15 mm deep.

Control   Maintain a maximum temperature of 90°F (32°C). High fertilization will also promote notching.

Selected Reference

Conover, C. A., and R. T. Poole. 1973. Factors influencing notching and necrosis of *Dracaena deremensis* 'Warneckii' foliage. Proc. Trop. Reg. Am. Soc. Hortic. Sci. 17:378-384.

Phyllosticta leaf spot

Figure 186

Cause   *Phyllosticta dracaenae*

Signs and symptoms   Spots are circular to slightly irregular and range from pinpoint to 1 cm in diameter. They appear mainly on the older leaves of plants and are usually tan with purple borders and yellow halos. Under conditions of high disease pressure, the spots may run together and the entire leaf may die.

Control   Use the same controls as those mentioned for anthracnose.

Selected References

Desai, M. V., and K. P. Patel. 1961. Leaf blight of *Dracaena* incited by *Phyllosticta draconis*. Plant Dis. Rep. 45:203.

Seymour, C. P. 1974. Phyllosticta leaf spot of *Dracaena*. Fla. Dept. Agric. & Cons. Serv. Plant Pathol. Circ. No. 143.

Pseudomonas leaf spot

Figure 187

Cause   *Pseudomonas* sp.

Signs and symptoms   Dark green, water-soaked spots gradually enlarge to 2 cm or more. They turn light brown with age and may have a necrotic center with a wide, water-soaked margin.

Control   Use the same controls as those listed for Erwinia blight.

Selected Reference

Miller, J. W., and C. Wehlburg. 1969. Bacterial leaf spot of *Dracaena sanderiana*. Proc. Fla. State Hortic. Soc. 82:368-370.

Pythium root rot and decline

Figure 188

Cause   *Pythium graminicola*

Signs and symptoms   Root rot is typified by wilting of plants and yellowing of lower leaves. The roots themselves are brown to black, reduced in mass, and mushy. The outer portion of infected roots can easily be pulled away from the inner core.

Control   Using pathogen-free potting medium and pots and growing plants on raised benches can eliminate much of this problem. If fungicides are needed, drenches with thiophanate methyl plus etridiazole or etridiazole alone can aid in control of Pythium root rot. Since other pathogens may be involved as well as *Pythium graminicola*, accurate diagnosis of the cause must be made prior to choice of fungicides.

Selected Reference

Uchida, J. Y., C. Y. Kadooka, and M. Aragaki. 1992. *Dracaena* decline and root rot. University of Hawaii at Manoa. College of Trop. Agric. & Human Res. HITAHR Brief No. 103.

Epipremnum (Pothos)

Pothos are climbing vines, native to Southeast Asia, commonly used as pedestal plants, on totems, in hanging baskets, or as ground covers in mass plantings. Plants are produced under 3,000 to 5,000 ft-c with temperatures between 70 and 90°F (21 and 32°C). Pothos tolerate indoor light levels of 50 to 100 ft-c. The most common diseases of pothos are bacterial and fungal stem and root rots, although these diseases can be easily avoided if pathogen-free plants and potting media are used. Mealybugs and caterpillars are the most serious pests; mites infestations are occasionally a problem.

Anthracnose

Figure 189

Cause   *Colletotrichum* spp.

Signs and symptoms   Anthracnose is characterized by yellow and later dark brown spots anywhere on the leaf. Yellowish masses of spores form in zones along leaf veins or in concentric rings in the spot. Spots are usually found in mist propagation or in the plant centers, where humidity and water levels are highest.

Control   Do not use any cuttings that have spots when taken from the stock plants. On rooted plants, minimize overhead irrigation and exposure to rainfall if possible. Fungicides that are effective for controlling anthracnose on pothos include mancozeb, thiophanate methyl, and iprodione.

Selected Reference

Graham, S. O., and J. W. Strobel. 1958. The incidence of anthracnose fungi on ornamental foliage plants in Washington State greenhouses. Plant Dis. Rep. 42:1294-1296.

Bird's nest fungi

Figure 190

Cause *Crucibulum, Cyathus, Nidula,* and *Sphaerobolus* spp.

Signs and symptoms Lower leaves of plants become speckled with black, tan, or gray circular structures that are easily mistaken for scale insects. The structures, which can be removed by scraping with a fingernail across the leaf surface, are the spores of the bird's nest fungus. They are very hard to smash, unlike the bodies of most scale insects.

Control Bird's nest fungi grow on wood materials such as benches, bark in potting media, and chips under benches. Avoid bringing the fungi into your operation by looking for the characteristic fruiting structures (bird's nests about 4 to 8 mm wide) on all wood materials. Noting which side of the plants is speckled can point out the actual source of the problem, since the structures are forcibly shot directly outward and stick to any surface. Chemical controls are rarely effective, and none are recommended. Bird's nest fungi cause cosmetic damage only; they do not cause a plant disease.

Selected References

Birchfield, W., J. L. Smith, A. Martinez, and E. P. Matherly. 1957. Chinese evergreen plants rejected because of glebal masses of *Sphaerobolus stellatus* on foliage. Plant Dis. Rep. 41:537-539.

Simone, G. W. 1987. The scale problem that isn't. Interior Landscape Industry 4(4):60, 61, and 64.

Chilling injury

Figure 191

Cause Exposure to temperatures below 50°F (10°C)

Signs and symptoms The leaves have black or brown dead areas, especially in the lighter portions (white or yellow), which develop 1 to 4 weeks after their exposure to cold.

Control Maintain production environment at 50°F (10°C) or above. Prevent plant exposure to cold air or cold water (condensate or drip-through from the roof of the structure). Low air temperatures account for most of the injury in open stock beds, while cold water draining through perforations in the polyethylene film used for lining shade houses can occur. Use solid-cover structures that are properly heated.

Selected Reference

Chase, A. R., and R. T. Poole. 1991. Effect of potassium rate, temperature and light on growth of pothos. University of Florida, Central Florida Research and Education Center-Apopka, CFREC-A Research Report, RH-91-11.

Erwinia leaf spot

Figure 192

Cause *Erwinia carotovora* and *E. chrysanthemi*

Signs and symptoms Bacterial leaf spot diseases are characterized by rapidly spreading, water-soaked lesions that form anywhere on the leaves. Under wet conditions, the centers of these spots may fall out. Sometimes leaf spots have a yellow border. A mushy soft rot of the lower end of a cutting is common. Sometimes the plants have a fishy, rotten odor, characteristic of *Erwinia* infections.

Control Bacterial leaf spot can be controlled by keeping water off the leaves. Choice of clean cuttings and strict sanitation are the most important control measures. During rooting of cuttings, applications of antibiotics such as streptomycin sulfate have aided in control.

Selected References

Chase, A. R. 1988. Controlling Erwinia cutting rot of marble queen pothos. University of Florida, Central Florida Research and Education Center-Apopka, CFREC-A Research Report, RH-88-4.

Knauss, J. F., and J. W. Miller. 1972. Description and control of the rapid decay of *Scindapsus aureus* incited by *Erwinia carotovora*. Proc. Fla. State Hortic. Soc. 85:348-352.

Manganese toxicity

Figure 193

Cause High pH and low manganese

Signs and symptoms Spotting or mottling on both upper and lower leaf surfaces, especially on older bottom leaves, occurs. Spots are purplish or tan and surrounded by a yellow halo.

Control Maintain the pH of the potting medium above 6.0 to minimize availability of excess manganese. Potting medium pH can be affected by the fertilizer source, irrigation water, and potting medium components. Dolomitic lime as an amendment can increase pH.

Selected Reference

Steinkamp, R. 1994. Avoid leaf mottling in pothos. Greenhouse Grower 12(5):106-107.

Nitrogen toxicity

Figure 194

Cause Excess nitrogen

Signs and symptoms Leaves are smaller than normal and have tan, greasy-appearing splotches within the light-colored areas. They may be lighter green than optimal, and desired variegation patterns may disappear.

Control Reducing the amount of fertilizer applied and leaching the potting medium are recommended for this disorder. In addition, plants may be transplanted to new potting media and left unfertilized until normal growth returns.

Sometimes damage is so severe that affected plants must be discarded. Recommended rates of 5 lb of N/1,000 ft²/month from a 3-1-2 (N-P-K) fertilizer should not be exceeded.

Selected Reference

Poole, R. T., A. R. Chase, and L. S. Osborne. 1991. Pothos. University of Florida, Central Florida Research and Education Center-Apopka, CFREC-A Foliage Plant Research Note, RH-1991-29.

Potassium deficiency

Figure 195

Cause   Insufficient potassium

Signs and symptoms   Leaves are mottled with yellowish spots that are translucent when viewed from below. They become progressively smaller as the condition worsens, and growth may cease. Severe symptoms may take from 4 to 6 months to appear and are therefore not as common on production plants as on stock plants.

Control   Regular applications of a balanced fertilizer (3-1-2, N-P-K) will prevent potassium deficiency.

Selected References

Chase, A. R., and R. T. Poole. 1991. Effect of potassium rate, temperature and light on growth of pothos. University of Florida, Central Florida Research and Education Center-Apopka, CFREC-A Research Report, RH-91-11.

Chase, A. R., and R. T. Poole. 1991. Effect of potassium and potting medium on growth of golden pothos. University of Florida, Central Florida Research and Education Center-Apopka, CFREC-A Research Report, RH-91-14.

Pseudomonas leaf spot

Figure 196

Cause   *Pseudomonas cichorii*

Signs and symptoms   Spots are small, translucent, and sometimes surrounded by a yellow halo. They become tan and can dry out as they age. *Aglaonema, Anthurium, Caladium, Monstera,* some *Philodendron* species, pothos, and *Xanthosoma* are susceptible to this pathogen.

Control   The only successful way to control this disease is to eradicate symptomatic plants. This should be done during the warm months when Pseudomonas leaf spot is most likely to appear. Water plants early in the day, use fans, and increase plant spacing to minimize this disease. Use of infected plants that are not showing symptoms (asymptomatic) generally results in cutting loss, since the bacterium is found inside the plant stem (systemic) and becomes active during rooting. Antibiotic and copper compounds may provide limited control of the leaf spot symptom. Keep plant foliage dry to minimize new infections. Most other foliage plants are susceptible to this bacterium and must be considered potential sources of infection.

Selected Reference

Wehlburg, C., C. P. Seymour, and R. E. Stall. 1966. Leaf spot of araceae caused by *Pseudomonas cichorii* (Swingle) Stapp. Proc. Fla. State Hortic. Soc. 79:433-436.

Pythium root rot

Figures 197 and 198

Cause   *Pythium splendens*

Signs and symptoms   Cuttings usually show poor rooting and have yellow leaves. Examination of the stem and roots reveals a mushy, black rot extending from the cut end into the upper portions of the stem and leaves. Root and stem rot usually occurs in patches on a propagation bench where it spreads into uninfected cuttings.

Control   Reducing water applications to the minimum level for good rooting also reduces root and stem rot disease. Always use disease-free propagation material, sterilized potting media, and raised benches when possible. Treatment with etridiazole, metalaxyl, or fosetyl aluminum can be very effective in control of Pythium root rot on pothos.

Selected References

Chase, A. R. 1982. Influence of soil drench fungicides on rooting of two foliage plants. University of Florida, Agricultural Research Center-Apopka, ARC-A Research Report, RH-82-12.

Chase, A. R. 1984. Some new soil fungicides for control of Pythium root rot of foliage plants. University of Florida, Agricultural Research and Education Center-Apopka, AREC-A Research Report, RH-84-23.

Chase, A. R. 1989. 1989 fungicide trials for control of *Cylindrocladium, Helminthosporium, Pythium* and *Rhizoctonia* diseases of ornamentals. University of Florida, Central Florida Research and Education Center-Apopka, CFREC-A Research Report, RH-89-15.

Chase, A. R., D. D. Brunk, and B. L. Tepper. 1985. Fosetyl aluminum fungicide for controlling Pythium root rot of foliage plants. Proc. Fla. State Hortic. Soc. 98:119-122.

Chase, A. R., and R. T. Poole. 1985. Irrigation frequency and Subdue rate affect Pythium root rot of pothos. University of Florida, Agricultural Research and Education Center-Apopka, AREC-A Research Report, RH-85-14.

Knauss, J. F. 1972. Description and control of Pythium root rot on two foliage species. Plant Dis. Rep. 56:211-215.

Rhizoctonia foot rot

Figure 199

Cause   *Rhizoctonia solani*

Signs and symptoms   Spots start on either petioles or leaf centers. They are brown and irregularly shaped. A mass of brownish mycelium covers the infected plants. Growth of mycelium from the potting medium onto the plant can escape notice and give the appearance that plants have been infected from an aerial source of inoculum. Close examination, however, generally reveals the presence of mycelium on stems prior to development of obvious symptoms. *Rhizoctonia* mycelia are usually reddish brown and have the consistency of a spider web.

Control Use the cultural controls listed for Pythium root rot. Disease is most severe when temperatures are between 68 and 86°F (20 and 27°C). Chemical control of this disease is excellent with PCNB, triflumizole, or thiophanate methyl and mancozeb.

Selected References

Chase, A. R. 1990. 1990 Fungicide trials for control of *Alternaria, Helminthosporium, Phytophthora* and *Rhizoctonia* diseases of ornamentals. University of Florida, Central Florida Research and Education Center-Apopka, CFREC-A Research Report, RH-90-25.

Chase, A. R. 1991. Characterization of *Rhizoctonia* species isolated from ornamentals in Florida. Plant Dis. 75:234-238.

Chase, A. R. 1992. Efficacy of thiophanate methyl fungicides for diseases of Florida ornamentals. Proc. Fla. State Hortic. Soc. 105: 182-186.

Chase, A. R., and T. A. Mellich. 1992. Controlling *Rhizoctonia* diseases on ornamentals with fungicides. University of Florida, Central Florida Research and Education Center-Apopka, CFREC-A Research Report, RH-92-8.

Millikan, D. F., and J. E. Smith, Jr. 1955. Foot rot of pothos, a disease caused by *Rhizoctonia*. Plant Dis. Rep. 39:240-241.

Southern blight

Figure 200

Cause *Sclerotium rolfsii*

Signs and symptoms The pathogen attacks all portions of the plant but is most commonly found on stems and leaves. Initially, symptoms on stems are confined to water-soaked, necrotic lesions at or near the soil line. White, relatively coarse mycelium grows in a fanlike pattern and may be seen on the soil surface, leaves, or stems. The round sclerotia form almost anywhere on the affected portions of the plant or the soil surface. They are initially white and cottony and approximately the size of a mustard seed. As sclerotia mature, they turn tan and eventually dark brown and harden. Mycelia and sclerotia generally develop concurrently with stem rot and wilting, allowing an accurate diagnosis of the problem. A cutting rot can develop on contaminated plant materials during the summer months.

Control This disease can be avoided by using proper cultural methods, including examining plant materials as they are brought into the nursery and growing plants on raised benches with new or sterilized pots and potting media. PCNB provides the most effective control of this disease but can result in stunting of plant tops and/or roots if it is used more than once per crop or at rates that are higher than recommended.

Xanthomonas blight

Figure 201

Cause *Xanthomonas campestris* pv. *dieffenbachiae*

Signs and symptoms Xanthomonas blight has been found on satin pothos but is not common. Most infections are confined to the leaf margins, but under conditions of high moisture and warm temperatures, infections of broad areas within the leaf blade are also found. These lesions are typically confined to areas between leaf veins. Sometimes lesions are small, water-soaked specks that enlarge into irregularly shaped areas.

Control Using raised benches and minimizing wetting of foliage are two of the most important cultural methods of controlling this disease. Maintain excellent sanitation, and exclude the pathogen from stock areas. Chemical control studies on other plants indicate that cupric hydroxide or streptomycin sulfate treatments are somewhat effective.

Episcia

Flame violets are native to tropical America and grow as ground covers or in hanging baskets. They are produced under 2,000 to 2,500 ft-c. Night temperatures must be kept above 65°F (18°C) or these plants rapidly decline and may die. Episcias are relatively disease-free, although mealybugs and mites are common pests.

Fusarium stem rot

Figure 202

Cause *Fusarium* spp.

Signs and symptoms Fusarium stem rot and blight typically appears as a soft, mushy rot at the base of a cutting or rooted plant. The pathogen sometimes forms tiny, bright red fruiting bodies or dusty tan masses of spores at the stem base of severely infected plants.

Control Remove infected plants from stock areas as soon as they are detected. Since Fusarium stem rot is similar in appearance to other diseases, accurate disease diagnosis is very important prior to applications of any fungicide. Always use pathogen-free cuttings and new or sterilized pots and potting media, and grow plants on raised benches with adequate spacing to allow rapid leaf drying after irrigation. No fungicide testing has been performed for Fusarium stem rot on flame violet.

Rhizoctonia aerial blight

Figure 203

Cause *Rhizoctonia solani*

Signs and symptoms A mass of brownish mycelia covers the infected plants. Growth of mycelia from the potting medium onto the plant can escape notice and give the appearance that plants have been infected from an aerial source of inoculum. Close examination, however, generally reveals the presence of mycelia on stems prior to development of obvious leaf and stem necroses. *Rhizoctonia* mycelia are usually reddish brown and have the consistency of a spider web.

Control Use the same controls as listed for Fusarium stem rot and blight.

Euphorbia

Most species of *Euphorbia* have succulent leaves and a milky sap and resemble cacti in some respects. The most common euphorb is poinsettia (*E. pulcherrima*). Many other euphorbs are used in small dish gardens, as desk-top plants, or even as large specimen plants, depending upon the species. Light levels of 3,000 to 5,000 ft-c are used for production, with a minimum night temperature of 55°F (13°C). Depending upon the species, euphorbs are hosts of numerous viruses, bacteria, and fungi that cause root and stem diseases. Pests of euphorbs include mites, scales, and mealybugs.

Erwinia soft rot

Figure 204

Cause   *Erwinia* spp.

Signs and symptoms   A blackened, wet, slimy lesion generally starts at the soil line at the base of the plant and progresses to the stem tip. Plants may wilt, become completely mushy, collapse, and often die.

Control   Remove and destroy infected plants as soon as they are found. Keep watering to a minimum, and avoid splashing, since this can spread the bacterium to other plants. Irrigate early in the day to allow rapid drying of the foliage, thereby reducing the ability of the bacterium to infect. Be sure to obtain an accurate diagnosis of the problem, since several of the diseases of euphorbs have very similar symptoms.

Selected Reference

Suslow, T., and A. H. McCain. 1979. Etiology, host range and control of a soft rot bacterium from cactus. (Abstr.) Phytopathology 69:921.

Fusarium cutting rot

Figure 205

Cause   *Fusarium* spp.

Signs and symptoms   Fusarium root and stem rot typically appears as a soft, mushy rot at the base of a cutting or rooted plant. The rotten area frequently contains the tan to orange masses of spores of the pathogen. Roots are mushy and brown and easily disintegrate when handled.

Control   Use the controls listed for Erwinia soft rot.

Myrothecium cutting rot and leaf spot

Figure 206

Cause   *Myrothecium roridum*

Signs and symptoms   Lesions generally appear at cutting bases or on leaves and result in poor rooting or even loss of cuttings. Necrotic areas are dark brown or black and mushy. Examination of the rotted areas reveals fruiting bodies (sporodochia), which are irregularly shaped and black and have a white fringe of mycelium.

Control   Using fungicides when temperatures are between 70 and 85°F (21 and 30°C) and fertilizing at recommended levels contribute to minimizing severity of *Myrothecium* diseases. Chlorothalonil, mancozeb, and iprodione each have been effective for *Myrothecium* control on many foliage plants.

Phomopsis stem spot

Figure 207

Cause   *Phomopsis* sp.

Signs and symptoms   Spots are sunken, tan, and irregularly shaped. In severe infections, the stem can become covered with these lesions and growth will be distorted.

Control   Minimize overhead irrigation and exposure to rainfall to reduce spread of the disease. Water early in the day, and use fans and plant spacing to facilitate rapid drying of the plants. Always use plants without disease symptoms for propagation. No information is available on chemical control of *Phomopsis* on euphorbs.

Phytophthora stem rot

Figure 208

Cause   *Phytophthora* spp.

Signs and symptoms   Root rots caused by *Phytophthora* spp. start as brownish root tips that rapidly disintegrate and cause the upper portions of the plant to yellow and wilt. The lower leaves of badly rotted cuttings drop off, and the bases of these cuttings are black or brown and mushy. This disease is common in poorly aerated, waterlogged soils.

Control   Always use clean or new pots and potting medium to reduce the chances of introducing pathogens into the production area. Growing plants away from the native soil is also a good idea, since the pathogens can be transferred readily to your euphorb crop. Broadly labeled products such as fosetyl aluminum or metalaxyl will be useful.

Rhizoctonia cutting rot

Figure 209

Cause   *Rhizoctonia solani*

Signs and symptoms   A brownish rot starts on cuttings at the soil line. It is also common for leaves to become infected. Under conditions of high temperatures and high moisture levels, the mycelium of the pathogen covers the infected cutting. *Rhizoctonia* mycelia are usually reddish brown and have the consistency of a spider web.

Control Use the same controls as listed for Phytophthora stem rot. You can apply drenches of thiophanate methyl if fungicides are needed.

Rhizopus blight

Figure 210

Cause *Rhizopus stolonifera*

Signs and symptoms A soft, mushy, brown rot can start anywhere on infected plants, including cutting bases, leaves, flowers, and shoot tips. The white mycelium and black sporangia of the pathogen form rapidly on all infected plant parts, giving them a fuzzy or bearded look.

Control This disease can be spread by air movement as well as by splashing from rainfall or irrigation practices. It is generally a problem under conditions of high temperatures and relative humidities. Keep plant stress as low as feasible to aid in resisting this disease. Extensive and conscientiously applied cultural controls have proven effective in controlling this disease on some floricultural crops. There are no fungicides labeled for this use, although dicloran is used in postharvest control of a similar disease on stone fruits.

Selected Reference

Mulrean, E. N., and A. H. McCain. 1979. A soft rot of *Euphorbia trigona* caused by *Rhizopus stolonifera*. (Abstr.) Phytopathology 69:1039.

Sunscald

Figure 211

Cause High light levels

Signs and symptoms Leaves rapidly develop tan to white, bleached patches. The spots can appear overnight after plants have been exposed to light much brighter than that to which they are accustomed.

Control Never move a plant from a low-light or shaded condition into very bright or direct sunlight. If plants must be moved to higher light, make the move in increments from low to medium to high light over a period of weeks to allow the leaves to adjust to the higher light levels.

Xanthomonas leaf spot

Figure 212

Cause *Xanthomonas campestris* pv. *poinsettiicola*

Signs and symptoms Symptoms are generally confined to pinpoint, yellow to tan lesions scattered across the leaf surface, although they can become large and confined between leaf veins. Lesions are mostly 2 mm wide with irregularly raised edges. Severe infections can cause distortion of new leaves as well as complete chlorosis and abscission of older leaves.

Control Eliminate all stock plants with Xanthomonas leaf spot. The disease is very difficult to control unless plants are produced without overhead watering or exposure to rainfall. Poinsettia and crotons are also susceptible to this disease.

Selected Reference

Miller, J. W., and C. P. Seymour. 1972. A comparative study of *Corynebacterium poinsettiae* and *Xanthomonas poinsettiaecola* on poinsettia and crown-of-thorns. Proc. Fla. State Hortic. Soc. 85: 344-347.

Ficus

Ficus spp. vary from vines to trees in their native habitats throughout the tropical regions of the world. Their interiorscape uses include hanging baskets or large specimen trees. Light levels for production of ficus vary from 1,500 to 8,000 ft-c depending upon the species involved. Many growers produce this plant in full sun and then move it into the appropriate shade level for acclimatization to the interior environment. During the past 10 years, a very serious disease of ficus has occurred on plants that are under stress indoors. This fungal disease does not damage the plants until they are indoors under stress from low light or other factors and always results in death of the tree. Although ficus require 150 ft-c for good growth in the interiorscape, they will tolerate 100 ft-c. Ficus are hosts of mites, mealybugs, scales, and thrips.

Anthracnose

Figure 213

Cause *Glomerella cingulata, Colletotrichum* spp.

Signs and symptoms Anthracnose is characterized by yellow and later dark brown spots anywhere on the leaf. Yellowish masses of spores form in zones along leaf veins or in concentric rings in the spot. Eventually, leaves may abscise. *Ficus elastica* cultivars are commonly infected with this pathogen during the summer months and appear especially susceptible when they are being rooted under mist conditions. *Glomerella* (sexual stage) and *Colletotrichum* (asexual stage) are often found in the same crop causing this disease.

Control Keep plant stresses from water and heat to a minimum. Do not use any cuttings that have spots when taken from the stock plants. On rooted plants, minimize overhead irrigation and exposure to rainfall if possible. Fungicides that are effective for controlling anthracnose include mancozeb, chlorothalonil, and iprodione.

Selected References

Agrawal, S. C., and S. B. Saksena. 1972. Experimental studies on leaf anthracnose of *Ficus elastica* Roxb. Current Science 41(7): 246-249.

Graham, S. O., and J. W. Strobel. 1958. The incidence of anthracnose fungi on ornamental foliage plants in Washington State greenhouses. Plant Dis. Rep. 42:1294-1296.

Bendiocarb phytotoxicity

Figure 214

Cause   Application of bendiocarb to the potting medium

Signs and symptoms   A single application of bendiocarb to the potting medium results in necrotic spots, tip necrosis, leaf cupping, and distortion.

Control   Always follow pesticide labels for applications rates, intervals, and especially sites. Products meant to be applied to leaves are not always safe when applied to the roots via the potting medium.

Selected Reference

Osborne, L. S., and A. R. Chase. 1986. Phytotoxicity evaluations of Dycarb on selected foliage plants. University of Florida, Agricultural Research and Education Center-Apopka, AREC-A Research Report, RH-86-12.

Boron toxicity

Figure 215

Cause   Excessive amounts of boron in fertilizer

Signs and symptoms   Marginal and tip chlorosis is followed by necrosis. Spots are irregularly shaped and can be water soaked. They greatly resemble those caused by some plant pathogens and can also be confused with symptoms caused by fluoride toxicity.

Control   Never apply a fertilizer that is high in boron. Check potting media or soil for excessive amounts of boron prior to its use. If this type of spot appears in an interiorscape, it is more likely caused by nutrient toxicity than by a disease.

Selected References

Marlatt, R. B. 1978. Boron deficiency and toxicity symptoms in *Ficus elastica* 'Decora' and *Chrysalidocarpus lutescens*. HortScience 13:442-443.

Poole, R. T., and C. A. Conover. 1985. Boron and fluoride toxicity of foliage plants. University of Florida, Agricultural Research and Education Center-Apopka, AREC-A Research Report, RH-85-19.

Botrytis blight

Figure 216

Cause   *Botrytis cinerea*

Signs and symptoms   Gray mold (Botrytis blight) of rubber tree cuttings usually starts as tan to brown lesions with concentric rings of light and dark tissue. Large, tan to brown leaf spots with concentric rings are usually found between the leaf and sheath or on leaf tips. Infections occur readily on tissue within the leaf sheath, where conditions

for infection are frequently ideal. Under optimum conditions for disease development, the emerging leaf rots completely. Botrytis blight occurs primarily on *Ficus elastica* during cool periods of the year, especially on cuttings.

Control   Watch for Botrytis when the following conditions occur: low light, high humidity, poor air circulation, and warm days with cool nights. Increase air circulation with fans and irrigate early in the day to allow the most rapid drying of plant foliage. Iprodione or vinclozolin give excellent control of this disease.

Selected Reference

Alfieri, S. A., Jr. 1966. Gray mold disease of ficus. Fla. Dept. of Agric. and Cons. Serv., Div. of Plant Indus., Pl. Pathol. Circ. 45, 2 pp.

Bromine toxicity

Figure 217

Cause   Excess bromine

Signs and symptoms   The youngest leaves become etched with a yellow marking that may look like feeding damage caused by thrips. Distortion and sometimes leaf drop can also occur.

Control   Do not apply bromine compounds directly to the foliage of sensitive cultivars or species. If bromine is needed as a water amendment for algae or disease control, then be sure to use rates less than 55 ppm of bromine, since this rate has been shown to cause damage on *F. benjamina* (weeping fig). Rates of 25 ppm of bromine have been safe when used on weeping fig in a propagation mist.

Selected Reference

Chase, A. R. 1991. Control of some fungal diseases of ornamentals with Agribrom. University of Florida, Central Florida Research and Education Center-Apopka, CFREC-A Research Report, RH-91-3.

Cercospora leaf spot

Figure 218

Cause   *Cercospora* sp.

Signs and symptoms   Tiny and slightly raised, red or dark green spots appear on the lower leaf surfaces of *Ficus elastica*. Severely infected leaves turn yellow (as shown in Figure 218) and drop. Cercospora leaf spot of climbing fig is characterized by light brown, irregularly shaped lesions with dark brown borders and chlorotic halos. Lesions occur primarily on leaf tips and margins of older leaves.

Control   Use the cultural controls listed for anthracnose. Weekly applications of zineb gave good control of Cercospora leaf spot of India rubber tree experimentally, but extended use caused significant stunting. If fungicides are used, be sure to get good coverage of leaf undersides, where the spores of the pathogen are formed.

Selected References

Alfieri, S. A., Jr. 1985. Cercospora leaf spot of climbing fig. Fla. Dept. Agric. & Cons. Serv., Div. of Pl. Indus., Pl. Path. Circ. No. 274.

Marlatt, R. B. 1970. Isolation, inoculation, temperature relations and culture of a *Cercospora* pathogenic to *Ficus elastica* 'Decora'. Plant Dis. Rep. 54:199-202.

Marlatt, R. B. 1972. Inoculation, incubation and abscission of *Ficus elastica* foliage with *Cercospora* disease. Plant Dis. Rep. 56: 1091-1092.

Corynespora leaf spot

Figure 219

Cause   *Corynespora cassiicola*

Signs and symptoms   Small to large, reddish leaf spots appear on the most recently mature leaves, and leaf abscission is common in severe infections when leaf spots expand between the veins. The lesions turn tan and sometimes have a narrow, dark brown margin or chlorotic halo.

Control   This disease occurs on both green and variegated forms of *Ficus benjamina* and *F. nitida* but is more severe on the variegated cultivars. Keep fertilizer applications at recommended levels, and eliminate overhead water if possible. Mancozeb and chlorothalonil are very effective in controlling Corynespora leaf spot on variegated *F. benjamina*.

Selected References

Chase, A. R. 1983. Controlling Corynespora leaf spot of *Ficus benjamina variegata*. Foliage Digest 6(11):10.

Chase, A. R. 1984. Leaf spot of *Ficus benjamina* caused by *Corynespora cassiicola*. Plant Dis. 68:251.

Crown gall

Figure 220

Cause   *Agrobacterium tumefaciens*

Signs and symptoms   Slightly swollen areas on the stems, leaf veins, and even roots are initially apparent. These swollen areas enlarge and become corky. In cases of severe infection, they may enlarge and merge to create a very distorted stem or root mass. Galls may also form on the ends of cuttings or on stems where cuttings have been removed.

Control   Remove and destroy all plants found infected with the bacterium, and then sterilize any cutting tools used on them. Since a fungus is also known to cause galls on *Ficus*, an accurate disease diagnosis must be made.

Selected References

Levine, M. 1921. Studies on plant cancers. II. The behavior of crown gall on the rubber plant (*Ficus elastica*). Mycologia 13:1-11.

Levine, M. 1924. Studies on plant cancers. VI. Further studies on the behavior of crown gall on the rubber plant, *Ficus elastica*. Mycologia 16:24-29.

Foliar nematode

Figure 221

Cause   *Aphelenchoides besseyi*

Signs and symptoms   Leaf spots begin near the midvein on lower leaves and extend to the margin. They are usually rectangular. *F. elastica* is the *Ficus* species most commonly infected with foliar nematode.

Control   Infestation of *F. elastica* occurs through movement of nematodes from weeds to lower leaves. Mow weeds in field plantings to stop this movement. Remove and destroy all plants with foliar nematode infestations, and never use them to propagate new plants.

Selected References

Marlatt, R. B. 1966. *Ficus elastica* a host of *Aphelenchoides besseyi* in a subtropical climate. Plant Dis. Rep. 50:689-691.

Marlatt, R. B. 1970. Transmission of *Aphelenchoides besseyi* to *Ficus elastica* leaves via *Sporobolus poiretii* inflorescences. Phytopathology 60:543-544.

Marlatt, R. B. 1975. Control of foliar nematode, *Aphelenchoides besseyi* in *Ficus elastica* 'Decora'. Plant Dis. Rep. 59:287.

Frost injury

Figure 222

Cause   Exposure to frost

Signs and symptoms   Leaves develop slightly sunken, tan or white areas anywhere on their surfaces. They usually are apparent within a day or two after a frost.

Control   Protect *Ficus* from frost by growing it in an enclosed structure where temperatures can be controlled or with overhead irrigation when necessary.

Helminthosporium leaf spot

Figure 223

Cause   *Bipolaris* spp.

Signs and symptoms   Small to large, brown leaf spots appear on the youngest mature leaves, and leaf abscission is common in severe infections when leaf spots expand between the veins.

Control   Treat as any other leaf spot disease by eliminating overhead watering and exposure to rainfall if possible. Mancozeb and chlorothalonil are very effective in controlling Helminthosporium leaf spot on many plants.

Myrothecium leaf spot

Figures 224, 225, and 226

Cause   *Myrothecium roridum*

Signs and symptoms   Leaf spots are generally found at wounds, although it is common to find no obvious wound and very large (up to 3 cm) leaf spots. The spots are usually tan to brown and may have a bright yellow border. Black and white fruiting bodies of the pathogen form in concentric rings on the leaf underside.

Control Follow the same controls as listed for anthracnose. Temperatures between 70 and 85°F (21 and 30°C) are optimal for disease development.

Phomopsis dieback

Figures 227, 228, and 229

Cause *Phomopsis cinerescens*

Signs and symptoms Phomopsis dieback causes a slow decline typified by leaf loss, twig death, stem cankers, and finally death of the tree. This disease can sometimes be found during production of ficus but more frequently becomes a serious problem when the plants are moved indoors. The pathogen apparently enters the plant through wounds created by pruning or during transport or adverse weather conditions, such as freezing. During production, infections usually remain latent. The first expression of the infection may not occur until the plants have been installed in the interiorscape and are subjected to water or light stress. Sections of the infected branches show the black streaks of fungal tissue characteristic of Phomopsis dieback.

Control The only effective controls for this problem are to use high-quality ficus free of obvious wounds and twig death and to maintain plants under optimum conditions after installation. It is extremely important that the ficus be fully acclimatized for the interior environment, since only then can the stress of transition be minimized. This disease does not cause leaf spots on the ficus and should be identified by examining stems below dead areas for cankers and splits.

Selected References

Anderson, R. G., and J. R. Hartman. 1983. Phomopsis twig blight on weeping fir indoors: A case study. Foliage Digest 6(1):5-7.

Hudler, G. W. 1979. Ficus problems. New York State Flower Industries Bulletin 104:10.

Nenschop, K., J. P. Tewari, and E. W. Toop. 1985. Phomopsis twig die-back of some woody interior ornamentals in Alberta. Foliage Digest 8(9):2-3.

Pseudomonas leaf spot

Figure 230

Cause *Pseudomonas cichorii*

Signs and symptoms Spots form on the recently matured leaves and start as water-soaked areas between veins. They enlarge and can expand along the vein and into the petiole. Spots are dark brown or black and usually do not have a yellow margin.

Control Eliminate all *F. lyrata* stock plants that have Pseudomonas leaf spot. The disease can be difficult to

control if plants are produced with overhead watering or exposure to rainfall and is more common during the cooler months. Bactericides such as copper-containing compounds may be somewhat effective if used on a preventative and regular basis, but none are labeled for use on *Ficus*.

Selected References

Chase, A. R. 1987. Leaf and petiole rot of *Ficus lyrata* cv. Compacta caused by *Pseudomonas cichorii*. Plant Pathol. 36:219-221.

Chase, A. R. 1988. Effect of fertilizer rate on growth of *Ficus lyrata* and susceptibility to *Pseudomonas cichorii*. HortScience 23:151-152.

Rhizoctonia aerial blight and root rot

Figure 231

Cause *Rhizoctonia solani*

Signs and symptoms Rhizoctonia aerial blight occurs primarily during the summer or warmer months. Brown, irregularly shaped spots form anywhere on the foliage but are most common within the crown of the plant, which is often wet. Sometimes the first symptoms form near the top of the plant, confusing the source of the disease (the soil). The disease spreads rapidly, and the entire plant may become covered with the brown, weblike mycelium of the pathogen.

Control A pathogen-free potting medium is the first step in the control of all soilborne pathogens. Plants should be produced from pathogen-free stock and grown in new or sterilized pots on raised benches. Since this pathogen inhabits the soil, both the roots and the foliage of the plants must be treated with a fungicide to provide optimal disease control. Chlorothalonil, iprodione, and thiophanate methyl are each effective in controlling this disease.

Selected References

Bolton, A. T. 1984. Root rot of *Ficus benjamina*. Plant Dis. 68:816-817.

Chase, A. R. 1991. Characterization of *Rhizoctonia* species isolated from ornamentals in Florida. Plant Dis. 75:234-238.

Southern blight

Figures 232 and 233

Cause *Sclerotium rolfsii*

Signs and symptoms Plants with southern blight may initially be similar in appearance to those infected with many other stem- or root-infecting fungi. As this disease advances, however, the white, cottony masses of mycelia and brown, seedlike sclerotia set it apart. The sclerotia usually form on the basal portion of stems of infected plants but may also be found on infected leaves. Eventually, the entire cutting or plant may be covered with the fungus.

Control Southern blight must be controlled through prevention with pathogen-free potting medium, pots, and planting materials. PCNB has been shown to be effective but may cause stunting if used more than once per crop at rates higher than recommended.

Selected Reference

West, E. 1947. *Sclerotium rolfsii* Sacc. and its perfect state on climbing fig. Phytopathology 37:67-69.

Xanthomonas leaf spot

Figure 234

Cause *Xanthomonas campestris* pv. *fici*

Signs and symptoms Foliar infections on ficus start as tiny, pinpoint, water-soaked areas that can rapidly enlarge. They tend to remain confined to the areas between leaf veins. Sometimes lesions have a bright yellow border. In severe infections, leaf abscission is common.

Control Eliminate all stock plants with Xanthomonas leaf spot. The disease can be difficult to control if plants are produced with overhead watering or exposure to rainfall. In addition, use of elevated rates of fertilizer has been shown to reduce the severity of Xanthomonas leaf spot on *Ficus benjamina*. There are large differences in susceptibility to this pathogen among *Ficus* species and cultivars (Table 12). Bactericides such as copper-containing compounds may be somewhat effective if used on a preventative and regular basis. Check labels for legal uses.

Selected References

Chase, A. R. 1988. Effect of fertilizer rate on susceptibility of *Ficus benjamina* to *Xanthomonas campestris* pv. *fici*. Proc. Fla. State Hortic. Soc. 101:339-340.

Chase, A. R. 1990. Effect of nitrogen and potassium on growth of *Ficus benjamina* and severity of Xanthomonas leaf spot. University of Florida, Central Florida Research and Education Center-Apopka, CFREC-A Research Report, RH-90-3.

Chase, A. R., and R. W. Henley. 1993. Susceptibility of some *Ficus* species to *Xanthomonas*. Southern Nursery Digest 27(6):20-21.

TABLE 12. Relative susceptibility of 19 *Ficus* species or cultivars to *Xanthomonas campestris* pv. *fici*

Species and/or cultivar	Disease severity
F. benjamina 'Phillipinense'	Severe
F. buxifolia	Severe
F. triangularis	Severe
F. benjamina 'Jacqueline'	Moderate to severe
F. benjamina 'Wintergreen'	Moderate to severe
F. benjamina 'Spearmint'	Moderate to severe
F. benjamina 'Green Mint'	Moderate to severe
F. benjamina 'Golden Green'	Moderate to severe
F. benjamina 'Nuda'	Moderate to severe
F. benjamina "common"	Moderate to severe
F. benjamina 'Exotica A'	Moderate to severe
F. benjamina 'Golden Princess'	Moderate to severe
F. benjamina 'Florida Spire'	Moderate to severe
F. mexicana	Moderate to severe
F. maclellandii 'Alii'	Moderate to severe
F. benjamina 'White Princess'	Moderate
F. retusa 'California Nitida'	Slight
F. 'Green Island'	Slight
F. retusa 'Green Gem'	Slight

Fittonia

Nerve plants are native to Colombia and Peru. They are commonly grown in dish gardens, terrariums, and hanging baskets. Production of nerve plants is optimal under 1,500 to 2,500 ft-c, with a night temperature no lower than 60°F (15°C). In the interiorscape, nerve plants should have light levels of 100 to 150 ft-c. Although nerve plants are subject to the same root diseases as most other foliage plants, they appear relatively disease free under most production systems. A bacterial disease caused by *Xanthomonas* is frequently misdiagnosed as a cultural disorder or pesticide phytotoxicity. Mealybugs and mites are the most common pests of these plants.

Bidens mottle

Figures 235 and 236

Cause Bidens mottle virus

Signs and symptoms Distortion of the normally symmetrical leaves is the most obvious symptom of this viral disease. Interveinal chlorosis and stunting of severely infected plants can occur as well. The disease appears to be most severe during the cool periods of the year but has rarely been seen on fittonias during the past 5 years.

Control This virus is transmitted from common weed hosts to the fittonia via aphid vectors. Remove weeds from around greenhouses as feasible, and keep aphid populations under control. Once plants are infected, they should be removed and destroyed, because they will not recover from the infection even if they do not show symptoms on a continuous basis.

Selected References

Zettler, F. W., J. A. A. Lima, and D. B. Zurawski. 1977. Bidens mottle virus infecting *Fittonia* spp. in Florida. Proc. Am. Phytopathol. Soc. 4:121-122.

Zurawski, D. B., D. E. Purcifull, and J. J. McRitchie. 1980. Bidens mottle virus of *Fittonia verschaffeltii*. Fla. Dept. Agric. & Cons. Serv. Plant Pathol. Circ. No. 215.

Chilling injury

Figure 237

Cause Air temperatures below 45 or 50°F (7 and 10°C)

Signs and symptoms The youngest leaves are the most sensitive to chilling injury and develop white to tan splotches, especially near the edges. Exposure to 50°F (10°C) will cause leaf necrosis but will not result in wilt or tip damage. Leaf wilt and flower collapse occur if nerve plants are exposed to temperatures lower than 50°F for extended periods of time.

Control Keep production air temperatures at least 55°F (13°C). The damage is permanent, but plants will produce healthy leaves when air temperatures are adequate unless the shoot tip has been damaged by extreme cold for extended periods.

Selected Reference

McConnell, D. B., D. L. Ingram, C. Groga-Bada, and T. J. Sheehan. 1982. Chilling injury of silver nerve plant. HortScience 17:819-820.

Rhizoctonia aerial blight

Figure 238

Cause   *Rhizoctonia solani*

Signs and symptoms   A mass of brownish mycelia covers the foliage of infected plants. Growth of mycelia from the potting medium onto the plant foliage can escape notice and give the appearance that plants have been infected from an aerial source. Close examination generally reveals the presence of the mycelia on plant stems prior to development of the aerial blight phase. *Rhizoctonia* mycelia are usually reddish brown and have the consistency of a spider web. Affected leaves and stems become brown and matted together.

Control   This disease is most severe during the summer months or at any time when the air temperatures are typically 80 to 90°F (27 to 32°C). Avoid applications of excess water to minimize conditions for disease development. Drench or foliar applications of thiophanate methyl may aid in disease control.

Selected Reference

Chase, A. R. 1991. Characterization of *Rhizoctonia* species isolated from ornamentals in Florida. Plant Dis. 75:234-238.

Xanthomonas leaf spot

Figure 239

Cause   *Xanthomonas campestris*

Signs and symptoms   Fittonias are commonly infected with this bacterium, although most producers do not recognize the symptoms as a disease problem. Marginal necrosis and vein necrosis are the most common symptoms. These are sometimes confused with irrigation, phytotoxicity, or temperature problems. All types of fittonias have been found susceptible to this pathogen, as well as some related plants such as *Aphelandra squarrosa* (zebra plant).

Control   Both streptomycin sulfate and copper compounds cause phytotoxicity on fittonias and can actually increase severity of Xanthomonas leaf spot because the bacteria invade areas damaged by the pesticides. The best way to control this disease is to establish plants that are free of the disease as propagative stock. All symptomatic plants should be collected and destroyed. Minimizing overhead irrigation will also reduce disease development and spread.

Selected Reference

Blake, J. H., A. R. Chase, and G. W. Simone. 1989. A foliar disease of *Fittonia verschaffeltii* caused by a pathovar of *Xanthomonas campestris*. Plant Dis. 73:269-272.

Hedera

English ivies are climbing vines native to North Africa, Asia, and southern Europe. Most English ivies are used in hanging baskets, on totems, or as ground covers in mass plantings. English ivies are primarily propagated from stem cuttings and are produced under 1,500 to 2,500 ft-c, with a minimum temperature of 55°F (13°C). Ivies require between 75 and 150 ft-c in the interiorscape to retain their quality. The most common disease of English ivy is a bacterial leaf spot, which has been reported from all areas throughout the world where English ivy is produced. Other diseases are caused by fungi that attack roots, stems, and leaves. English ivies host a multitude of pests including spider and tarsonemid (bud) mites, mealybugs, scales, aphids, and lepidopterous larvae.

Anthracnose

Figure 240

Cause   *Colletotrichum trichellum*

Signs and symptoms   Colletotrichum leaf spot (also called anthracnose) is very similar in appearance to Xanthomonas leaf spot. Sometimes lesions appear black with tiny, black specks in their centers. These specks are the fruiting structures of the pathogen. This disease is more commonly a problem on ivy in the landscape than during production. Diagnosis of these symptoms by culturing is recommended to ensure the appropriate choice of control methods.

Control   Reduce overhead watering as much as possible, since it is necessary for disease development and spread. Table 13 lists anthracnose resistance levels for some ivy

TABLE 13. Relative susceptibility[a] of *Hedera helix* (English ivy) cultivars to pathogens and phytotoxicity[b]

Cultivar	Xanthomonas	Colletotrichum	Streptomycin sulfate
Sweet Heart	L	–	L–M
Telecurl	L–M	–	L
Gold Dust	L	L	L–M
Eva	L	–	L
Green Feather	M	L	L
Ivalace	M	–	L
Perfection	L	M	L
Hahn variegated	M–H	M	L
Brokamp	H	–	H
California	L	–	M
Manda crested	L	H	H
Gold Heart	L–M	–	L–M

[a] L = low, M = moderate, H = high, and – = not tested.
[b] Adapted from Osborne, L. S., and A. R. Chase, 1985, Susceptibility of cultivars of English ivy to two-spotted spider mite and Xanthomonas leaf spot, HortScience 20:269-271; and Pierce, L., and A. H. McCain, 1983, Colletotrichum leaf spot of English ivy: Chemical control and cultivar susceptibility, Calif. Plant Pathol. 62:1-4.

cultivars. Several fungicides, including iprodione and thiophanate methyl, are effective for anthracnose control. Do not use chlorothalonil on ivy because necrosis and distortion of new leaves can occur occasionally on some cultivars.

Selected References

Garren, K. H. 1946. A disease of English ivy in Georgia. Plant Dis. Rep. 30:209-210.

Pierce, L., and A. H. McCain. 1983. Colletotrichum leaf spot of English ivy: Chemical control and cultivar susceptibility. Calif. Plant Pathol. 62:1-4.

Ridings, W. H., and S. A. Alfieri, Jr. 1973. Colletotrichum leaf spot of English ivy. Fla. Div. of Agr. and Cons. Serv. Plant Pathol. Circ. No. 131.

Botrytis blight

Figures 241 and 242

Cause   *Botrytis* sp. and *Sclerotinia* sp.

Signs and symptoms   Botrytis blight first appears as relatively large, grayish or tan areas on leaf margins and in their centers. Spots enlarge rapidly and can encompass the entire leaf. The dusty, grayish tan spores of the pathogen form readily in the dead tissue and can be easily seen with the naked eye. Affected leaves generally become covered with spores and collapse. At times, the perfect (sexual) stage of *Botrytis* forms instead and is responsible for the disease.

Control   Botrytis blight occurs during the winter months when days are cool and short and humidity in the greenhouses is high. Venting the greenhouse at sundown to reduce relative humidity and using fans and increased plant spacing to improve air circulation to facilitate leaf drying are recommended. Iprodione and vinclozolin provide good to excellent control of Botrytis blight.

Bromine toxicity

Figure 243

Cause   Excess bromine

Signs and symptoms   The youngest leaves become etched with a yellow marking that may look like feeding damage caused by thrips. Distortion and severe stunting of leaves can also occur.

Control   Do not apply bromine compounds directly to the foliage of sensitive cultivars or species. If bromine is needed as a water amendment for algae or disease control, then be sure to use rates less than 55 ppm of bromine, which has been shown to cause this damage on English ivy. Rates of 25 ppm bromine are safe when applied in a propagation mist.

Selected References

Chase, A. R. 1990. Control of some bacterial diseases of ornamentals with Agribrom. Proc. Fla. State Hortic. Soc. 103:192-193.

Chase, A. R. 1991. Control of some fungal diseases of ornamentals with Agribrom. University of Florida, Central Florida Research and Education Center-Apopka, CFREC-A Research Report, RH-91-3.

Chlorothalonil phytotoxicity

Figure 244

Cause   Application of chlorothalonil

Signs and symptoms   Immature leaves become distorted, and necrotic spots may form. Spots turn reddish brown with age.

Control   Do not use chlorothalonil on English ivy, since some cultivars are sensitive to this fungicide.

Selected Reference

Chase, A. R. 1989. 1989 fungicide trials for control of *Cylindrocladium, Helminthosporium, Pythium,* and *Rhizoctonia* diseases of ornamentals. University of Florida, Central Florida Research and Education Center-Apopka, CFREC-A Research Report, RH-89-15.

Fertilizer burn

Figure 245

Cause   Excessive soluble salts, especially from fertilizer

Signs and symptoms   Leaves and stems wilt and collapse. Marginal burn and tipburn on leaves are common. Roots on affected plants are sparse, darkened, and sometimes mushy.

Control   Follow recommendations for fertilizer application to English ivy. Use lower levels during slow growing periods and higher levels when plants are growing rapidly. Current recommendations are 2.5 to 3.0 lb N/1,000ft²/month from a 3-2-2 or 2-1-2 fertilizer. When 20-20-20 is used, 14.4 lb/1,000ft²/month is recommended.

Selected Reference

Henley, R. W., A. R. Chase, and L. S. Osborne. 1991. English ivy. Univ. of Florida, Central Florida Research and Education Center-Apopka, CFREC-A Foliage Plant Research Note, RH-91-15.

Fusarium blight

Figures 246 and 247

Cause   *Fusarium tricinctum* and others

Signs and symptoms   Infection of leaves under very wet conditions results in tan, papery leaf spots sometimes with concentric rings of light and dark tissue. This disease can also affect stems and roots, causing typical symptoms that do not allow diagnosis without culturing the pathogen.

Control   Remove infected plants from stock areas as soon as they are detected. Never use them for propagation. Since Fusarium blight is similar in appearance to several other diseases of English ivy, accurate disease diagnosis is very important prior to choice and application of pesticides. If stem rot or cutting rot is a problem, treatment of the cuttings with a dip or a post-sticking drench of thiophanate methyl should diminish losses. Foliar sprays with chlorothalonil, thiophanate methyl, or iprodione should aid in control of Fusarium leaf spot.

Selected Reference

El-Gholl, N. E., J. J. McRitchie, C. L. Schoulties, and W. H. Ridings. 1978. The identification, induction of perithecia, and pathogenicity of *Gibberella* (*Fusarium*) *tricincta* n. sp. Can. J. Bot. 56:2203-2206.

Pythium root rot

Figure 248

Cause   *Pythium* spp.

Signs and symptoms   Infected plants exhibit poor growth and color, and basal leaves turn brown and curl downward. Root rot causes stems to wilt. Roots themselves are discolored, mushy, and sparse and disintegrate readily when handled.

Control   Cultural controls include the use of pathogen-free potting media, pots, and plant material as well as minimal water applications. Always use a potting medium that does not hold water for extended periods of time. Chemical controls that are generally effective include fosetyl aluminum, etridiazole, and metalaxyl.

Rhizoctonia aerial blight

Figure 249

Cause   *Rhizoctonia solani*

Signs and symptoms   Rhizoctonia blight occurs primarily during the hot summer months when humidities are very high in the growing area. Disease development is rapid and can occur in less than 1 week if conditions are optimal. Brown, irregularly shaped lesions form all over the plant. Although the first symptoms sometimes appear on the top of the plant, the pathogen inhabits the potting medium and is not as readily spread by air movement as many other fungal leaf spot organisms. The lesions spread rapidly, and the reddish brown, weblike mycelium can cover the entire plant.

Control   Cultural control of this disease is the same as that listed for Pythium root rot. Thiophanate methyl is somewhat effective as a soil drench for control of Rhizoctonia root rot on ivy. Fungicides that provide control of Rhizoctonia blight include vinclozolin, iprodione, and thiophanate methyl.

Selected References

Chase, A. R. 1989. 1989 fungicide trials for control of *Cylindrocladium, Helminthosporium, Pythium,* and *Rhizoctonia* diseases of ornamentals. University of Florida, Central Florida Research and Education Center-Apopka, CFREC-A Research Report, RH-89-15.

Chase, A. R. 1991. Characterization of *Rhizoctonia* species isolated from ornamentals in Florida. Plant Dis. 75:234-238.

Streptomycin sulfate phytotoxicity

Figure 250

Cause   Application of streptomycin sulfate

Signs and symptoms   Yellowing and sometimes whitening of new leaves occur on ivy. Sometimes leaves are also stunted. This is especially apparent on solid green cultivars but also occurs on variegated cultivars.

Control   Although streptomycin sulfate is effective in controlling bacterial leaf spot on English ivy, it is not recommended for use on ivy because of its phytotoxicity. Plants treated with as little as 100 ppm a single time develop severe chlorosis. Table 13 lists cultivar sensitivity to streptomycin sulfate.

Xanthomonas leaf spot

Figure 251

Cause   *Xanthomonas campestris* pv. *hederae*

Signs and symptoms   Xanthomonas leaf spot of ivy cultivars is characterized by brown to black, circular to irregularly shaped spots found first on the oldest foliage. Many times the spots have a bright yellow halo or margin and a water-soaked edge. Infection of immature leaves results in speckling and deformity of these leaves. Older leaves usually develop circular to irregularly shaped, black, necrotic lesions with bright yellow halos visible on both leaf surfaces.

Control   Use pathogen-free stock plants, minimize overhead watering, and remove infected plants to decrease spread to adjacent plants. Fertilizer applied at rates higher than recommended has been shown to reduce susceptibility of ivy to *X. campestris* pv. *hederae*. Other plants in the Araliaceae family such as sch022effleras and aralias are also susceptible. Table 13 lists responses of some ivy cultivars to Xanthomonas leaf spot. Up to 90% control has been possible with cupric hydroxide, fosetyl aluminum, or a 1% solution of white vinegar. Using chemical applications, minimizing overhead irrigation, growing resistant cultivars where possible, and managing host nutrition are recommended for control of Xanthomonas leaf spot on members of the Araliaceae family.

Selected References

Chase, A. R. 1989. Effect of fertilizer rate on susceptibility of *Hedera helix* to *Xanthomonas campestris* pv. *hederae*, 1988. Biol. & Cult. Tests 4:82.

Chase, A. R. 1989. Nitrogen source and rate affect severity of Xanthomonas leaf spot of *Hedera helix*. University of Florida, Central Florida Research and Education Center-Apopka, CFREC-A Research Report, RH-89-1.

Chase, A. R. 1989. Effect of Osmocote rate on severity of Xanthomonas leaf spot of English ivy. University of Florida, Central Florida Research and Education Center-Apopka, CFREC-A Research Report, RH-89-11.

Dye, D. W. 1967. Bacterial spot of ivy caused by *Xanthomonas hederae* (Arnaud. 1920) Dowson, 1939, in New Zealand. New Zealand J. Sci. 10:481-485.

Osborne, L. S., and A. R. Chase. 1985. Susceptibility of cultivars of English ivy to two-spotted spider mite and Xanthomonas leaf spot. HortScience 20:269-271.

Hoya (Wax plant)

Hoya originated in Southeast Asia and the Pacific islands. They are used primarily in hanging baskets. They grow best when temperatures are maintained above 60°F (15°C). *Hoya* are very slow growing and tolerant of indoor conditions and require relatively low light. Their waxy leaves, stems, and flowers continue to maintain an attractive appearance even under the low humidities typical of indoor environments. *Hoya* are relatively disease free but are frequently infested with scale insects or mealybugs.

Botrytis blight

Figure 252

Cause *Botrytis cinerea*

Signs and symptoms Botrytis blight usually appears on lower leaves of cuttings in contact with the potting medium. The water-soaked lesion may enlarge rapidly to encompass a large portion of the leaf blade or even the entire cutting. The area turns necrotic and dark brown to black with age. When night conditions are cool, day conditions warm, and moisture conditions high, the pathogen readily sporulates on both leaves and flowers, covering them with grayish brown, dusty masses of conidia. Cuttings rooted during the winter are especially susceptible to Botrytis blight, since the environment is ideal for the disease and very poor for rapid growth of the plant.

Control Controlling Botrytis blight is particularly important during the winter months. Cultural methods that improve foliage drying and reduce moisture condensation on foliage during the nights reduce Botrytis blight. Both iprodione and vinclozolin have been shown to be effective for control of Botrytis blight. Check labels for legal use on *Hoya*.

Myrothecium leaf spot

Figure 253

Cause *Myrothecium roridum*

Signs and symptoms Lesions generally appear at the edges and tips of leaves and at broken leaf veins. Necrotic areas are dark brown and initially appear water soaked. Examination of the bottom leaf surface generally reveals sporodochia, which are irregularly shaped and black and have a white fringe of mycelium. Sporodochia form in concentric rings within the necrotic areas.

Control Using fungicides when temperatures are between 70 and 85°F (21 and 30°C), minimizing wounding, and fertilizing at recommended levels contribute to minimizing severity of Myrothecium leaf spot of foliage plants. Chlorothalonil and iprodione have been effective experimentally for *Myrothecium* control on other foliage plants.

Selected Reference

Chase, A. R. 1983. Influence of host plant and isolate source on Myrothecium leaf spot on foliage plants. Plant Dis. 67:668-671.

Leea

Leea originated in tropical Africa, Asia, and Australia. They are small shrubs and trees used mainly as specimen plants in the interiorscape. Plants are propagated from seeds and cuttings. Diseases include a few stem rots and leaf spots.

Calonectria collar rot and leaf spot

Figure 254

Cause *Calonectria crotalariae*

Signs and symptoms Calonectria collar rot and leaf spot causes wilting and rotting of the basal portion of the stem and crown. The rot is typically black and covered with the orange red perithecia of the pathogen. Leaf spots of naturally infected plants are blackish brown and circular (5 to 10 mm in diameter).

Control Cultural controls include use of pathogen-free potting media, pots, and plant material as well as minimal water applications.

Selected Reference

Ko, W. H., J. Y. Uchida, R. K. Kunimoto, and M. Aragaki. 1981. Collar rot and leaf spot of *Leea* caused by *Calonectria crotalariae*. Plant Dis. 65:621.

Phytophthora blight

Figures 255 and 256

Cause *Phytophthora* sp.

Signs and symptoms Infected plants exhibit poor growth and color, and basal leaves turn brown and curl downward. Root rot sometimes occurs, although leaf spot and stem rot appear to be more common. Leaf spots are large, gray to black, and water soaked.

Control The pathogen that causes this problem continues to grow within the plant even when environmental conditions are not favorable. Prevention is the key to disease control. Use pathogen-free seedlings, pots, and potting media. The same pathogen also causes damping-off of *Leea* seedlings, so it is important to have each disease diagnosed and discard all flats with the disease. Experimentally, metalaxyl has provided the best disease control.

Selected Reference

Aragaki, M., J. Y. Uchida, P. S. Yahata, and C. Y. Kadooka. 1990. Foliar blight of leea caused by a species of *Phytophthora*. Hawaii Inst. Trop. Agric. and Human Resources. Univ. Hawaii at Manoa. HITAHR Brief No. 093.

Xanthomonas leaf spot

Figure 257

Cause *Xanthomonas campestris* pv. undescribed

Symptoms Small, water-soaked spots form anywhere on young leaves. They become darkened with age and reach 5 mm wide. This disease is common on plants grown under overhead irrigation or in the landscape. The pathovar of *X. campestris* that causes this disease has not been fully described.

Control Use pathogen-free stock plants, minimize overhead watering, and rogue infected plants to decrease spread to adjacent plants. The host range of this pathovar of *X. campestris* is apparently restricted to *Leea*.

Selected Reference

Patel, A. M. 1969. Bacterial blight of *Leea edgewirthii* incited by *Xanthomonas leeanum*, sp. nov. Current Sci. 21:519-520.

Maranta

Prayer plants are herbaceous plants native to tropical America. They are grown under much the same conditions as calatheas in both production and interior uses and suffer from many of the same pests and diseases. Light levels up to 3,500 ft-c are recommended for production. In addition to the pathogens and pests listed for *Calathea*, *Maranta* are also subject to cucumber mosaic virus, which dramatically marks their foliage.

Benomyl toxicity

Figure 258

Cause Application of benomyl as a drench at the recommended rate to red maranta

Signs and symptoms Leaves turn yellow and sometimes white and are reduced in size.

Control Do not apply benomyl to red maranta. A phytotoxic reaction does not always occur, but one cannot predict when it will be a problem or when the compound can be used safely.

Selected References

Chase, A. R., and L. S. Osborne. 1981. Pesticide evaluations in support of registrations on tropical foliage plants. University of Florida, Agricultural Research Center-Apopka, ARC-A Research Report, RH-81-14.

Chase, A. R., and G. W. Simone. 1985. Phytotoxicity on foliage ornamentals caused by bactericides and fungicides. Univ. of Florida. Pl. Pathol. Fact PP-30.

Bleach toxicity

Figure 259

Cause Direct sprays of bleach solutions to leaves

Signs and symptoms Marantas grown in ground beds are sometimes accidentally sprayed with dilute (1:10) bleach solutions. Although they are sometimes unaffected, at other times leaves develop a whitened appearance. Damaged leaves are especially apparent along the outer portions of the beds near the walkways.

Control Try not to spray any foliage plants with bleach or other solutions that are being used to clean walks, benches, or other inanimate surfaces.

Cucumber mosaic

Figures 260 and 261

Cause Cucumber mosaic virus (CMV)

Signs and symptoms CMV causes dramatic symptoms on both red and green maranta. Leaves may be slightly distorted and reduced in size, but the most obvious symptom of CMV infection is the bright yellow patterns formed on the leaves. These patterns are generally jagged and alternate with the normal coloration of the affected leaf.

Control Although the symptoms of CMV are so striking, there is no evidence that the damage caused is other than aesthetic. The only recommended control is removal of plant material showing these symptoms. Propagation of material with CMV will transfer the virus to new plants.

Selected Reference

Hearon, S. S. 1979. A ringspot of prayer plant caused by a strain of cucumber mosaic virus. Plant Dis. Rep. 63:32-36.

Helminthosporium leaf spot

Figure 262

Cause *Drechslera setariae*

Signs and symptoms Helminthosporium leaf spot is frequently a problem on maranta produced in ground beds where plants stay very moist for long periods of time. The spots first appear as tiny, water-soaked areas that turn yellow and finally brown. Spots are normally very small (1 mm wide or less) and give the affected leaves a speckled appearance. In severe cases, spots merge and form large (up to 1 cm), irregularly shaped areas that are tan with a yellow edge.

Control Minimizing the period of time leaves are wet can dramatically reduce disease severity. Eliminate overhead watering, or apply water early in the day to allow rapid drying of foliage. Plants that are watered in the late afternoon may remain wet for the entire night, which promotes disease. Chlorothalonil and iprodione have been found effective for control of this disease experimentally.

Selected References

Chase, A. R. 1986. Current developments in chemical control of foliage plant diseases. Proc. of the 2nd conference on insect and disease management on ornamentals. Soc. Am. Florists 2:130-137.

Chase, A. R. 1986. Efficacy of Chipco 26019 flowable and wettable powder formulations for control of Alternaria and Drechslera leaf spots. University of Florida, Agricultural Research and Education Center-Apopka, AREC-A Research Report, RH-86-2.

Simone, G. W., and D. D. Brunk. 1983. New leaf spot disease of *Calathea* and *Maranta* spp. incited by *Drechslera setariae*. Plant Dis. 67:1160-1161.

Pythium root rot

Figure 263

Cause   *Pythium* spp.

Signs and symptoms   Leaves wilt, may turn yellow or pale green, and eventually die. Plants are frequently stunted, and examination of roots reveals their rotted condition. Initial infections of the roots appear as small, water-soaked, grayish or brown areas. These spots can rapidly expand to affect the entire root system. Severely infected plants may have no living roots remaining by the time they are examined.

Control   Prevention is always the best control of a soilborne pathogen like *Pythium*. Use clean pots and potting media, and grow plants on raised benches. Avoid overwatering, since waterlogged roots are easily attacked by *Pythium* spp. Etridiazole is effective for Pythium root rot control on maranta.

Pellionia and *Pilea*

Pellionias are woody or herbaceous plants native to tropical and subtropical Asia. They are used as small specimen plants, in hanging baskets, and in dish gardens or terrariums. Plants should be produced with light levels of 1,000 to 2,000 ft-c and temperatures between 65 and 85°F (18 and 30°C). Pellionias should have 75 to 100 ft-c in the interior. Both fungal and bacterial diseases can occur on pellionias. Mites, mealybugs, and scales are the most common pests of these plants. Snails and slugs also feed on pellionias.

Pileas are herbaceous plants native to tropical and subtropical areas. Some are upright in growth, while others are trailing and used in hanging baskets. Small dish gardens and terrariums commonly include a pilea. *Pilea cadierei*, the aluminum plant, requires interior light levels of 75 to 150 ft-c. Production conditions and pests and pathogens of pileas are similar to those listed for pellionias.

Edema

Figure 264

Cause   Cool air conditions and excess soil moisture

Signs and symptoms   Leaves develop a warty appearance on the undersides, especially near the edges. These raised areas are easiest to see on the undersides of the leaves and may be slightly water soaked or tan.

Control   Be careful to control the amount of water supplied to plants when the air temperatures are unseasonably cool.

Myrothecium leaf spot

Figure 265

Cause   *Myrothecium roridum*

Signs and symptoms   Brown to black, circular lesions form on leaf margins and centers. The lesions may have concentric rings of light and dark tissue, or they may be water soaked and uniformly black. The lower surface of the lesion frequently has the fungal fruiting bodies, which are irregularly shaped and black and have a white fringe around the borders.

Control   Young plants are highly susceptible to this disease and may be lost if precautions are not taken. Disease is most severe during periods of the year when air temperatures are between 60 and 85°F (15 and 30°C). Little, if any, disease occurs at other times. Use pathogen-free cuttings, avoid wounding plants, and do not fertilize in excess of recommended rates. Many other hosts of *M. roridum* exist, including aglaonema, dieffenbachia, ficus, philodendron, and syngonium. Fungicides that have aided in control under experimental conditions include mancozeb, chlorothalonil, iprodione, and captan. Check labels for use on these plants.

Selected Reference

Chase, A. R. 1983. Influence of host plants and isolate source on Myrothecium leaf spot of foliage plants. Plant Dis. 67:668-671.

Pythium root rot

Figure 266

Cause   *Pythium* spp.

Signs and symptoms   Pythium root rot is most often caused by *P. splendens*. Plants appear stunted, chlorotic, and wilted, even when soil moisture is high. Roots are usually rotted, brown to black, and mushy. The outer portion of the root tissue (cortex) is easily pulled away from the inner core, leaving a fine, hairlike root system when plants are removed from the potting medium. Since these symptoms can be caused by many different soilborne pathogens, an accurate diagnosis can be made only when the roots are tested for these organisms.

Control Cultural control of soilborne diseases is based on the use of pathogen-free seedlings or cuttings, pots, and potting media. Grow plants on raised benches away from the native soil, since it can be a source of infections or can become contaminated and infect future crops. Drench applications of several fungicides are effective in controlling Pythium root rot (e.g., etridiazole, fosetyl aluminum, and metalaxyl). Be sure to check labels prior to applying these compounds, since not all are registered for use on *Pellionia* and *Pilea* spp.

Rhizoctonia aerial blight

Figure 267

Cause *Rhizoctonia solani*

Signs and symptoms Discrete lesions that may coalesce to form blotchy, dead areas appear all over the plant foliage. The spider web-like mycelium of the pathogen develops all over the aerial portions of the plant and can cover infected plants completely. Affected tissue wilts and turns necrotic rapidly. The mycelium of *Rhizoctonia* is usually tan to reddish brown.

Control Cultural controls for Rhizoctonia aerial blight are the same as those mentioned for Pythium root rot, since both are soilborne pathogenic fungi. Chemical control of this disease can be achieved with applications of thiophanate methyl.

Selected Reference

Chase, A. R. 1991. Characterization of *Rhizoctonia* species isolated from ornamentals in Florida. Plant Dis. 75:234-238.

Xanthomonas leaf spot

Figures 268 and 269

Cause *Xanthomonas campestris* pv. unnamed

Signs and symptoms Symptoms on aluminum plant (*Pilea cadierei*) are dry, tan, irregularly shaped lesions found primarily in the white areas of the leaves. In advanced infections, the lesions fall away, leaving ragged holes in the leaf blades that can appear similar to insect feeding damage. Lesions on other pileas tend to be dark brown to black and may be angular, since they rarely spread across leaf veins. Symptoms on pellionias are dryish, irregularly shaped lesions with corky, slightly raised borders found mainly on leaf undersides. Hosts include aluminum plant, creeping charley, *Pilea spruceana*, *Pilea invulcrata* (silver tree), satin pellionia, and trailing watermelon begonia. Creeping charley and satin pellionia are most susceptible, while trailing watermelon begonia is relatively resistant.

Control Chemical control of this disease is rarely successful, and cultural methods should be the first line of defense. Elimination of overhead watering and/or exposure to rainfall aid in control of disease development and spread. However, once infection occurs, lesions can expand even when leaves are kept dry. Discard all plants infected with this pathogen, and never use infected plants for stock, since the disease is easily carried on tissue even when no symptoms are evident. Unfortunately, plants with the highest quality were found most susceptible to this disease in fertilizer trials. Other plants susceptible to this bacterium include *Ficus* spp. and bird-of-paradise. Chemical control of this disease on aluminum plant, trailing watermelon begonia, and satin pellionia was achieved with preventive applications of streptomycin sulfate or a combination of cupric hydroxide and mancozeb.

Selected References

Chase, A. R. 1989. Effect of fertilizer level on severity of Xanthomonas leaf spot of *Pilea spruceana*. J. Environ. Hort. 7:47-49.

Chase, A. R., and J. B. Jones. 1987. Leaf spot and blight of *Strelitzia reginae* (bird-of-paradise) caused by *Xanthomonas campestris*. Plant Dis. 71:845-847.

Miller, J. W., and A. R. Chase. 1986. New disease of *Pellionia* and *Pilea* species caused by *Xanthomonas campestris*. Plant Dis. 70: 346-348.

Peperomia

Peperomias are small, herbaceous plants native to tropical and subtropical regions all over the world. They are used in dish gardens and terrariums and as small, desk-top plants. Light levels of 1,500 to 3,500 ft-c are recommended for plant production; interiorscape light levels of 75 to 150 ft-c are acceptable. Many fungal pathogens occur on peperomias and cause root, stem, and leaf diseases. Bud mites, thrips, and mealybugs are common pests.

Anthracnose

Figure 270

Cause *Colletotrichum* spp.

Signs and symptoms Anthracnose is characterized by yellow and later dark brown spots anywhere on the leaf. Yellowish masses of spores form in zones along the leaf veins or in concentric rings in the spot. Eventually leaves may abscise. Peperomias are especially susceptible when they are being rooted under mist conditions.

Control Keep plant stresses from water and heat to a minimum. Do not use any cuttings that have spots when taken from the stock plants. On rooted plants, minimize overhead irrigation and exposure to rainfall if possible. Fungicides effective for controlling anthracnose include mancozeb, chlorothalonil, and iprodione.

Cercospora leaf spot

Figure 271

Cause *Cercospora* sp.

Signs and symptoms Cercospora leaf spot is typified by tan to black, raised areas found on leaf undersides. The areas are similar in appearance to a condition called edema and are swollen and irregularly shaped. *Peperomia obtusifolia* cultivars are especially susceptible to Cercospora leaf spot.

Control Controls listed for anthracnose should be effective for Cercospora leaf spot as well. Thiophanate methyl has been effective for control of this disease on other ornamentals. Be sure to spray the undersides of the leaves, where the spores are produced.

Selected Reference

Alfieri, S. A., Jr. 1968. Cercospora and edema of *Peperomia*. Proc. Fla. State Hortic. Soc. 81:388-391.

Myrothecium leaf spot

Figure 272

Cause *Myrothecium roridum*

Signs and symptoms Myrothecium leaf spot most frequently appears on wounded areas of leaves, such as injured tips and breaks in the main vein that occur during handling. The leaf spots are watery and nearly always contain the black and white fungal fruiting bodies in concentric rings near the outer edge of the leaf undersides.

Control Avoid wounding leaves, and keep the foliage as dry as possible. Many other plants are hosts of *M. roridum*, including *Aglaonema, Aphelandra, Begonia, Calathea, Dieffenbachia, Spathiphyllum,* and *Syngonium*, and these plants must be included in control programs. Iprodione and mancozeb provide good control of this disease on other foliage plants.

Selected Reference

Chase, A. R. 1983. Influence of host plant and isolate source on Myrothecium leaf spot on foliage plants. Plant Dis. 67:668-671.

Phyllosticta leaf spot

Figure 273

Cause *Phyllosticta* sp.

Signs and symptoms Phyllosticta leaf spot occurs primarily on the watermelon peperomia. Leaf spots are dark brown to black and dryish and have concentric rings of light and dark tissue. Spots commonly start on leaf margins and can spread across the entire leaf.

Control Remove and destroy infected leaves. Keep foliage dry, and treat with thiophanate methyl or mancozeb according to labeled directions. This disease was not affected by fertilizer rate in experimental tests.

Selected References

Chase, A. R. 1988. Effect of fertilizer level on severity of Phyllosticta leaf spot of *Peperomia obtusifolia*, 1987. Biol. & Cult. Tests 3:82.
Chase, A. R. 1987. Phyllosticta leaf spot of *Peperomia* species and cultivars in Florida. University of Florida, Central Florida Research and Education Center-Apopka, CFREC-A Research Report, RH-87-3.

Phytophthora stem and crown rot

Figure 274

Cause *Phytophthora parasitica*

Signs and symptoms Plants rot at the soil line and show a mushy, black lesion that can extend upward into the leaves. Entire plants can be lost, and cuttings are especially prone to this disease.

Control Always use new or disinfested pots and potting medium, and grow plants on raised benches to avoid infection from the native soil. Avoid overwatering plants, since this makes diseases caused by *Phytophthora* and *Pythium* spp. more severe. Metalaxyl and several formulations of etridiazole are effective in controlling these diseases.

Selected References

Ark, P. A., and T. A. DeWolfe. 1951. Phytophthora rot of *Peperomia*. Plant Dis. Rep. 35:46-47.
Humphreys-Jones, D. R. 1980. *Phytophthora nicotianae* var. *nicotianae* on *Peperomia magnoliaefolia* and *Kalanchoe blossfeldiana*. Plant Pathol. 29:98-99.
Siradhana, B. S., C. W. Ellett, and A. F. Schmitthenner. 1968. Pathogenic and cultural variation in *Phytophthora nicotianae* var. *parasitica* from greenhouse plants. Phytopathology 58:718-719.
Siradhana, B. S., C. W. Ellett, and A. F. Schmitthenner. 1968. Crown rot of *Peperomia*. Plant Dis. Rep. 52:244.

Pythium root rot

Figure 275

Cause *Pythium splendens* and many other *Pythium* species

Signs and symptoms Roots of infected plants are sparse, blackened, and mushy, and their outer cortex can be easily removed from the inner core. Stunting is common and is sometimes the only obvious symptom of infection.

Control Use the controls listed for Phytophthora stem and crown rot. In addition, research has shown roots of underfertilized peperomias to be more susceptible to Pythium root rot than those of well-fertilized or even overfertilized plants.

Selected References

Chase, A. R., D. D. Brunk, and B. L. Tepper. 1985. Fosetyl aluminum fungicide for controlling Pythium root rot of foliage plants. Proc. Fla. State Hortic. Soc. 98:119-122.
Chase, A. R., and D. E. Munnecke. 1978. Pythium root rot and stunting of *Peperomia obtusifolia* var. *variegata*. Plant Dis. Rep. 62:314-315.
Chase, A. R., and R. T. Poole. 1984. Investigations into the roles of fertilizer level and irrigation frequency on growth, quality and severity of Pythium root rot of *Peperomia obtusifolia*. J. Am. Soc. Hortic. Sci. 109:619-622.

Kidney, B. A. 1979. Host range, virulence and control of *Pythium splendens* Braun from *Peperomia orba* Bunt. Proc. Fla. State Hortic. Soc. 92:355-358.

Rhizoctonia leaf spot

Figure 276

Cause   *Rhizoctonia* sp.

Signs and symptoms   *Peperomia* species are susceptible to *Rhizoctonia* sp., which causes a mushy, dark brown to black leaf spot. Spots are elliptical to irregularly shaped; concentric rings of raised and depressed tissue form in these areas. Under warm conditions, the weblike mycelium of the pathogen can be seen covering the affected plant.

Control   Always use pathogen-free cuttings and new or sterilized pots and potting media, and grow plants on raised benches. Remove and destroy severely infected plants or areas in the stock bed. Treat the bed with thiophanate methyl, being sure to thoroughly saturate the potting medium as well as to cover the tops of the plants, since this pathogen is soilborne.

Selected References

Chase, A. R. 1991. Characterization of *Rhizoctonia* species isolated from ornamentals in Florida. Plant Dis. 75:234-238.

Munnecke, D. E., and P. A. Chandler. 1953. Some diseases of variegated peperomia. Plant Dis. Rep. 37:434-435.

Ring spot

Figure 277

Cause   Cucumber mosaic virus

Signs and symptoms   Infected plants show a variety of symptoms including ring spots (rings of light or dark pigmentation), leaf distortion, and stunting for the green variety of *P. obtusifolia*. The virus appears as necrotic lesions (brown areas) on the variegated cultivars. Infected leaves generally fall off the plant.

Control   Collect and destroy all peperomias with these symptoms, since no chemicals can control a viral disease. Be careful not to transmit the virus by using contaminated cutting tools; clean tools between plants if this viral disease is suspected.

Selected References

Corbett, M. K. 1956. Virus ring spot of *Peperomia obtusifolia* and *Peperomia obtusifolia* var. *variegata*. Proc. Fla. State Hortic. Soc. 69:357-360.

Creager, D. B. 1941. Ring spot of popular peperomias caused by virus. Florists' Rev. 87(2256):15-16.

Southern blight

Figure 278

Cause   *Sclerotium rolfsii*

Signs and symptoms   Stem rot caused by this pathogen is characterized by a brown, mushy area at the soil line of the cutting. Plants in the rooting phase as well as established plants are frequently lost to this disease. The brown resting bodies of the pathogen are commonly found in the rotted area. These structures are tan to dark brown, round, and the size of mustard seeds. Masses of white, cottony mycelial growth are also found.

Control   Cuttings should be inspected carefully for symptoms of this disease and discarded if they are infected. Always use new potting medium and pots, and watch plants carefully for symptoms of stem rot. PCNB is the only effective chemical control for southern blight and should be used only according to the label, because it can cause stunting when used too often on the same crop.

Selected Reference

Alfieri, S. A., Jr., and J. F. Knauss. 1972. Stem and leaf rot of *Peperomia* incited by *Sclerotium rolfsii*. Proc. Fla. State Hortic. Soc. 85:352-357.

Philodendron

Philodendrons make up one of the most popular and common groups of foliage plants produced today. They are both vining and upright in habit and are native to tropical America. Interiorscape use of these plants ranges from hanging baskets to ground covers to specimen plants. Philodendrons are produced with 1,500 to 6,000 ft-c or full sun (*Philodendron selloum*). Under interiorscape conditions, many philodendrons tolerate 75 to 150 ft-c, although *P. selloum* requires more light than other species. Philodendrons are hosts of many pathogens, including serious bacterial and fungal leaf, stem, and root diseases. Pests of philodendrons include mealybugs and occasionally mites and aphids.

Bird's nest fungi

Figures 279 and 280

Cause   *Crucibulum* spp. and others such as shotgun fungi

Signs and symptoms   The fungus usually grows on decaying wood or compost in potting medium, mulched walks, or ground beds. When the fungus is mature, the spores are produced and shot onto all nearby surfaces, including plant leaves. Since they grow below the plants, most of the spores are found on leaf undersides, where they can be confused with scale insects or weed seeds such as *Oxalis*. They stick firmly to the surfaces on which they land and are very hard to remove from plant leaves without damaging the leaves themselves.

Control   Although a variety of fungicides have been tried for control of this type of fungus, removing the wood product that supports growth of shotgun fungi is still the best control.

Botrytis blight

Figure 281

Cause   *Botrytis cinerea*

Signs and symptoms   Botrytis leaf blight usually occurs on lower leaves of cuttings in contact with the potting medium. The water-soaked lesion may enlarge rapidly to encompass a large portion of the leaf blade or even the entire cutting. The infection may start as a small, white spot but can become necrotic and dark brown to black with age. When night conditions are cool, day conditions warm, and moisture conditions high, the pathogen readily sporulates on leaves, covering them with grayish brown, dusty masses of conidia.

Control   Methods that improve foliage drying and reduce moisture condensation on foliage during the night reduce the need for fungicide application. Iprodione and vinclozolin are effective for *Botrytis* control on philodendron.

Dactylaria leaf spot

Figure 282

Cause   *Dactylaria humicola*

Signs and symptoms   Initial symptoms of infection occur on both surfaces of young leaves as tiny, water-soaked areas. Lesions turn yellow to tan as they mature and commonly have depressed centers. These symptoms are easily confused with those caused by thrips feeding. Lesions are more numerous on the lower leaf surface, and only immature leaves of heart-leaf philodendron are susceptible to the pathogen. *Philodendron scandens* subsp. *oxycardium, P. hastatum, P. selloum,* and *P. pedatum* (= *P. laciniatum*) are hosts of the fungus. Disease occurs primarily during the summer months on ground bed plantings of the host.

Control   Increased air circulation obtained by raising plants off the ground has reduced occurrence of many foliar diseases of this crop. This also allows for better coverage by fungicides. Fungicides such as chlorothalonil and mancozeb provide excellent control of this disease on heart-leaf philodendron.

Selected Reference

Knauss, J. F., and S. A. Alfieri, Jr. 1970. Dactylaria leaf spot, a new disease of *Philodendron oxycardium* Schott. Proc. Fla. State Hortic. Soc. 83:441-444.

Dasheen mosaic

Figure 283

Cause   Dasheen mosaic virus (DMV)

Signs and symptoms   Chlorotic streaking and mosaic patterns as well as distortion of new leaves are found in philodendron infected with DMV. Growth of infected plants is reduced compared with that of healthy plants, even when obvious symptoms are not present.

Control   This virus is also a pathogen of *Aglaonema, Dieffenbachia,* and *Spathiphyllum,* and control practices should include these genera as well. Removal and destruction of infected plants is the only way to stop the spread of the virus. Aphid control and periodic sterilization of cutting instruments are also important ways if minimizing virus spread, since the virus may be vectored by either method.

Selected References

Simone, G. W., and F. W. Zettler. 1990. Dasheen mosaic disease of araceous foliage plants. Univ. of Florida Coop. Exten. Plant Pathol. Fact Sheet PP-42.

Wisler, G. C., F. W. Zettler, R. D. Hartman, and J. J. McRitchie. 1978. Dasheen mosaic virus infections of philodendrons in Florida. Proc. Fla. State Hortic. Soc. 91:237-240.

Erwinia blight

Figures 284 and 285

Cause   *Erwinia chrysanthemi* and *E. carotovora* subsp. *carotovora*

Signs and symptoms   Erwinia blight appears initially as tiny, water-soaked areas, primarily on leaves, that expand rapidly into irregularly shaped tan to black areas. The bacterium spreads from leaf lesions into petioles, causing complete collapse of infected leaves. Leaf and petiole disintegration is characteristic for Erwinia blight and can occur as soon as 2 days after infection. Infected plants produced in an enclosed area, such as a greenhouse, have a characteristic unpleasant (rotten fish) odor. This disease is most severe under conditions of high moisture and temperature. The bacterium appears to be dormant during the cool winter months.

Control   Removal and destruction of infected tissue is most desirable, and infected plants should never be used as cutting sources. Considerable research has been conducted on this problem in an effort to identify a cultural means of control. Free moisture on leaf surfaces is needed for infection, but wounding of the tissue is not required. Unlike some other bacterial diseases, extended periods of misting are not necessary for infection to occur on *P. selloum,* although these conditions do increase disease severity. Disease severity also increases as temperature increases and is most severe on plants with either too little or too much fertilizer. Minimizing water applications, using the recommended rate of fertilizer, and spacing plants to allow rapid drying of leaves are recommended controls for Erwinia blight. The wide host range of these bacteria makes it necessary to control the disease on all susceptible plants to reduce spread from one crop to another. Preventative sprays of streptomycin sulfate may aid in control if leaves are kept as dry as possible. A combination spray of mancozeb and basic copper sulfate or cupric hydroxide may also be effective.

Selected References

Chase, A. R. 1983. Attempts to control Erwinia blight of *Philodendron selloum* with some unusual compounds. University of Florida, Agricultural Research Center-Apopka, ARC-A Research Report, RH-83-15.

Chase, A. R. 1986. Effect of experimental bactericides on three bacterial diseases of foliage plants. J. Environ. Hort. 4:37-41.

Haygood, R. A., and D. L. Strider. 1979. Influence of temperature, inoculum concentration and wounding on infection of *Philodendron selloum* by *Erwinia chrysanthemi*. Plant Dis. Rep. 63:578-580.

Haygood, R. A., and D. L. Strider. 1981. Influence of moisture and inoculum concentration on infection of *Philodendron selloum* by *Erwinia chrysanthemi*. Plant Dis. 65:727-728.

Haygood, R. A., D. L. Strider, and E. Echandi. 1982. Survival of *Erwinia chrysanthemi* in association with *Philodendron selloum*, other greenhouse ornamentals, and in potting media. Phytopathology 72:853-859.

Haygood, R. A., D. L. Strider, and P. V. Nelson. 1982. Influence of nitrogen and potassium on growth and bacterial leaf spot of *Philodendron selloum*. Plant Dis. 66:728-730.

Knauss, J. F., and J. W. Miller. 1974. Etiological aspects of bacterial blight in *Philodendron selloum* caused by *Erwinia chrysanthemi*. Phytopathology 64:1526-1528.

Heater damage

Figure 286

Cause   High temperature

Signs and symptoms   Leaves rapidly develop tan to white, irregularly shaped areas. The spots appear overnight after plants have been exposed to high temperatures during the winter months when heaters are in use.

Control   Try to position heaters farther away from plants to decrease the potential for damage. Certain types of shields may help to protect the plants closest to the heaters.

Phytophthora leaf spot

Figure 287

Cause   *Phytophthora parasitica*

Signs and symptoms   Lesions are dark brown, water soaked, irregularly shaped, and 1 to 3 cm wide. The disease is most severe during the summer months in ground beds of philodendron.

Control   Growing plants in new or sterilized potting media on raised benches eliminates much of the source of this disease. Preventive sprays of mancozeb or chlorothalonil have been shown to provide control. In addition, etridiazole, fosetyl aluminum, or metalaxyl are effective in controlling *Phytophthora* in the potting medium.

Pseudomonas leaf spot

Figure 288

Cause   *Pseudomonas cichorii*

Signs and symptoms   The symptoms of this disease are similar in appearance to those caused by Erwinia leaf spot, except that the lesions rarely become mushy and do not appear water soaked. The yellow border shown in Figure 288 is not characteristic of Pseudomonas leaf spot.

Control   Minimize foliar wetting from irrigation or rainfall, use only pathogen-free plants as stock, and remove diseased materials as soon as they are found. Bactericide control should be the same as that mentioned for Erwinia blight.

Selected Reference

Wehlburg, C., C. P. Seymour, and R. E. Stall. 1966. Leaf spot of araceae caused by *Pseudomonas cichorii* (Swingle) Stapp. Proc. Fla. State Hortic. Soc. 79:433-436.

Pythium root rot

Figure 289

Cause   *Pythium* spp.

Signs and symptoms   Pythium root rot of foliage plants is most often caused by *P. splendens*. Plants appear stunted, chlorotic, and wilted even when soil moisture is high. Roots are usually rotted, brown to black, and mushy. The outer portion of the root tissue (cortex) is easily pulled away from the inner core, leaving a fine, hairlike root system when plants are removed from the potting medium.

Control   Follow the same controls listed for Phytophthora blight.

Red-edge disease

Figure 290

Cause   *Xanthomonas campestris* pv. *dieffenbachiae*

Signs and symptoms   Xanthomonas disease of heart-leaf philodendron was named red-edge disease after the conspicuous red edge that typifies this disease on this plant. Most infections are confined to the margin, but under conditions of high moisture and warm temperatures, infections of broad areas within the leaf blade are also found. These lesions are typically confined to areas between the leaf veins. Sometimes lesions are also small, water-soaked specks, which enlarge into irregularly shaped areas.

Control   Control of Xanthomonas diseases of foliage plants has been based mainly upon sanitation and exclusion of the pathogen from stock areas. Using raised benches and minimizing wetting of foliage are two of the most important cultural controls of red-edge disease. Red-edge disease of heart-leaf philodendron decreased as nitrogen application increased. Neither potassium nor phosphorous level affected disease severity in experimental tests. Chemical control studies have shown both cupric hydroxide and streptomycin sulfate to be effective in disease control.

Selected References

Chase, A. R., and R. T. Poole. 1993. Effects of light level and nitrogen fertilization on growth of heart-leaf philodendron stock plants and severity of red-edge disease. University of Florida, Central Florida Research and Education Center-Apopka, CFREC-A Research Report, RH-93-9.

Harkness, R. W., and R. B. Marlatt. 1970. Effect of nitrogen, phosphorus and potassium on growth and *Xanthomonas* disease of *Philodendron oxycardium*. J. Am. Soc. Hortic. Sci. 95:37-41.

Knauss, J. F., W. E. Waters, and R. T. Poole. 1971. The evaluation of bactericides and bactericide combinations for the control of bacterial leaf spot and tip burn of *Philodendron oxycardium* incited by *Xanthomonas dieffenbachiae*. Proc. Fla. State Hortic. Soc. 84:423-428.

McFadden, L. A. 1967. A *Xanthomonas* infection of *Philodendron oxycardium* leaves. (Abstr.) Phytopathology 57:343.

Wehlburg, C. 1968. Bacterial leaf spot and tip burn of *Philodendron oxycardium* caused by *Xanthomonas dieffenbachiae*. Proc. Fla. State Hortic. Soc. 81:394-397.

Rhizoctonia aerial blight

Figure 291

Cause   *Rhizoctonia solani*

Signs and symptoms   Rhizoctonia aerial blight occurs primarily during the summer or warmer months. Disease development can occur in less than a week, so plants should be checked carefully and frequently. Brown, irregularly shaped spots form anywhere on the foliage but most commonly within the crown of the plant, which is often wet. Sometimes the first symptoms form near the top of plant, confusing the source of the disease (the soil). The disease spreads rapidly, and the entire plant can become covered with the brown, weblike mycelium of the pathogen.

Control   A pathogen-free potting medium is the first step in controlling all soilborne pathogens. Plants should be produced from pathogen-free stock and grown in new or sterilized pots on raised benches. Since this pathogen inhabits the soil, both the roots and the foliage of the plants must be treated with a fungicide to provide optimal disease control. A combination drench-spray will best accomplish this. Chlorothalonil, iprodione, and thiophanate methyl are effective in controlling this disease. Check labels for legal uses on philodendrons.

Selected Reference

Chase, A. R. 1991. Characterization of *Rhizoctonia* species isolated from ornamentals in Florida. Plant Dis. 75:234-238.

Southern blight

Figure 292

Cause   *Sclerotium rolfsii*

Signs and symptoms   Plants with southern blight may initially be similar in appearance to those infected with many other stem or root infecting fungi. As this disease advances, however, the white, cottony masses of mycelia and brown, seedlike sclerotia set it apart. The sclerotia usually form on the basal portions of stems of infected plants but may also be found on infected leaves. Eventually the entire cutting or plant may be covered with the fungus.

Control   Use pathogen-free potting medium, pots, and planting materials. PCNB has been shown to be effective; but because of the possibility of stunting, it should be used only when the disease has been properly diagnosed. Be sure to check the product label for use on philodendron.

Selected Reference

French, A. M., and C. W. Nichols. 1954. *Sclerotium rolfsii* on *Philodendron cordatum* in southern California. Plant Dis. Rep. 38:530.

Plectranthus (Swedish ivy)

Swedish ivies are native to tropical Asia, Africa, and Australia. They are most frequently used in hanging baskets. Light levels of 3,000 to 4,000 ft-c and temperatures above 55°F (13°C) produce good-quality plants. They can tolerate freezing temperatures. Mealybugs and whiteflies as well as a few diseases can cause problems.

Myrothecium leaf spot

Figure 293

Cause   *Myrothecium roridum*

Signs and symptoms   Lesions generally appear at the edges, tips, and broken veins of leaves. Necrotic areas are dark brown and initially appear water soaked. Examination of the bottom leaf surface generally reveals sporodochia, which are irregularly shaped and black, have a white fringe of mycelium, and form in concentric rings within the necrotic areas.

Control   Using fungicides when temperatures are between 70 and 85°F (21 and 27°C), minimizing wounding, and fertilizing at recommended levels lessen the severity of Myrothecium leaf spot of foliage plants. Chlorothalonil and iprodione have been effective for control of *Myrothecium* on other foliage plants.

Selected Reference

Chase, A. R. 1983. Influence of host plant and isolate source on Myrothecium leaf spot on foliage plants. Plant Dis. 67:668-671.

Pseudomonas leaf spot

Figure 294

Cause   *Pseudomonas cichorii*

Signs and symptoms   Spots are water soaked and turn dark green to black. They may have a yellow edge, but this is not common. There are rarely more than two spots on a leaf, although loss of the leaf often occurs if it is infected before it expands completely. Defoliation can occur as well.

Control Avoid overhead watering as much as possible to reduce conditions for infection and spread of the pathogen. Never use symptomatic plants for stock plants. Many other foliage plants are hosts of *P. cichorii,* including ficus, fern, aroids, and schefflera. Bactericide sprays are not effective in controlling this disease.

Selected Reference

Miller, J. W. 1991. Bacterial blight of Swedish ivy caused by *Pseudomonas cichorii.* Fla. Dept. Agric. & Cons. Serv. Div. of Plant Industry. Plant Pathology Circ. 341.

Rhizoctonia aerial blight

Figure 295

Cause *Rhizoctonia solani*

Signs and symptoms Rooting cuttings may be completely covered by a mass of brownish mycelia. Growth of mycelia from the potting medium onto larger plants can escape notice and give the appearance that plants have been infected from an aerial source of inoculum. Close examination, however, generally reveals the presence of mycelia on stems prior to development of obvious foliar symptoms. This disease is most common during the hottest times of the year, particularly when the plant foliage remains wet for long periods of time or relative humidity is high.

Control Use pathogen-free cuttings and new pots and potting media, and avoid extremes in soil moisture. Chemical control of diseases caused by *Rhizoctonia* has been investigated on many plants with a variety of fungicides. The fungicide most widely used for soil drench control of *Rhizoctonia* diseases is thiophanate methyl. Remember that the potting medium or soil is a common source of this soilborne plant pathogen.

Polypodiaceae (Ferns)

Asplenium

Bird's nest ferns are primarily grown in terrariums and dish gardens and as small, desk-top plants. Some specimens may reach 1 m in width and height. Bird's nest ferns are grown with light intensities between 2,000 and 4,000 ft-c and a minimum night temperature of 60°F (15°C). Indoors, these plants require 100 to 150 ft-c. Plants are susceptible to several bacteria that cause leaf blights as well as to the foliar nematode. Scales, lepidopterous larvae, and mealybugs are common pests.

Davallia

Rabbit's foot ferns are epiphytic natives of tropical and subtropical regions of Europe, Asia, and Africa. They are used in dish gardens and terrariums and sometimes for hanging baskets or small, desk-top plants. Light intensities of 2,000 to 2,500 ft-c are recommended for production, with a minimum temperature of 65°F (18°C). This fern requires at least 150 ft-c for good interiorscape maintenance. Rabbit's foot ferns are hosts of many of the same pests and pathogens as bird's nest fern.

Nephrolepis

There are many species and cultivars of Boston fern produced as foliage plants, most of which originated in the tropical and subtropical regions of the world. They are used indoors in mass plantings and hanging baskets and sometimes as table-top specimens. Plants should be grown under 1,500 to 3,000 ft-c and maintained with a minimum night temperature of 65°F (18°C). Indoors, Boston ferns require 75 to 150 ft-c. The most serious diseases of these plants are caused by soilborne fungi and root-infesting nematodes. Common pests include caterpillars, scales, and mealybugs.

Botrytis blight

Figure 296

Cause *Botrytis cinerea*

Signs and symptoms Spots usually appear on the leaf underside, especially on petioles near the pot rim or in contact with the potting medium. A small, water-soaked spot can rapidly enlarge and cover the entire leaf. Sporulation on necrotic leaves appears as a powdery, grayish brown mass.

Control Watch for *Botrytis* when the following conditions occur: low light, high humidity, poor air circulation, and warm days with cool nights. Increase air circulation with fans and irrigate early in the day to allow the most rapid drying of plant foliage. Check labels for products containing iprodione or vinclozolin for use on specific ferns, since these fungicides are especially effective in controlling Botrytis blight on many plants.

Drought

Figure 297

Cause Lack of water

Signs and symptoms Ferns have a gray cast and become stunted. Wilting may or may not occur.

Control Ferns will turn gray if they do not receive sufficient water. The growth rate and runner production will be decreased if potting medium is not continuously moist. Increase irrigation to supply sufficient water, and monitor the plant more carefully.

Foliar nematode

Figure 298

Cause *Aphelenchoides fragariae*

Signs and symptoms Spots caused by foliar nematodes are sometimes similar to those caused by the bacterial pathogens described earlier. Small, water-soaked spots form, generally near the frond base. Spots rapidly turn brown to black, and distortion of the fronds may occur if large areas are infected. Affected tissues remain turgid and do not collapse. Spread of the nematode within the leaves is usually inhibited by the large leaf veins, making spots somewhat angular (vein delimited).

Control The most effective control of foliar nematodes on ferns is destruction of infested plants. These nematodes easily contaminate pots and bench surfaces covered with organic matter as well as the potting medium. Sanitation between crops can greatly reduce nematode spread from one crop to the next. In addition, avoid producing plants in contact with the ground, since it can be a source of many plant-parasitic nematodes.

Selected References

Ark, P. A., and C. M. Tompkins. 1946. Leaf nematode infestation of bird's nest fern. Phytopathology 36:892-893.

Johnson, A. W., and D. L. Gill. 1975. Chemical control of foliar nematodes (*Aphelenchoides fragariae*) on 'Fluffy Ruffles' fern. Plant Dis. Rep. 59:772-774.

Stokes, D. E. 1966. Effects of *Aphelenchoides fragariae* on bird's-nest fern and azaleas. Proc. Fla. State Hortic. Soc. 79:436-438.

Stokes, D. E. 1967. Newly reported fern hosts of *Aphelenchoides fragariae* in Florida. Plant Dis. Rep. 51:508.

Herbicide damage

Figure 299

Cause Volatilization of an herbicide (e.g., alachlor, dichlobenil, metalochlor, napropamide, prometon, tebuthiuron, or trifluralin)

Signs and symptoms Marginal burning occurs on older leaves and complete necrosis (browning) on immature or young leaves.

Control Never use unlabeled herbicides in enclosed structures such as greenhouses. Applications of some herbicides to the ground under benches can damage crops by volatilization. Tests have shown that some herbicides remain active in the soil for long periods of time (a year or more) and can continue to cause significant crop damage.

Selected Reference

Poole, R. T. 1985. Weeds in the greenhouse. Foliage Digest 11(5):8.

Myrothecium leaf spot

Figure 300

Cause *Myrothecium roridum*

Signs and symptoms Myrothecium leaf spot most frequently appears on wounded areas of leaves such as breaks that occur during handling. The leaf spots are watery and nearly always contain the black and white fungal fruiting bodies in concentric rings near the outer edges. They are usually seen on the leaf undersides. Newly planted, tissue-cultured explants are especially susceptible to this disease.

Control Avoid wounding leaves, and keep the foliage as dry as possible. Avoid using plantlets or cuttings with symptoms of Myrothecium leaf spot. Temperatures between 70 and 85°F (21 and 30°C) are most favorable for disease development. Many other plants including *Aglaonema*, *Aphelandra*, *Begonia*, *Calathea*, *Spathiphyllum*, and *Syngonium* are hosts of *M. roridum*, and these plants must be included in control programs. Fungicidal control with chlorothalonil or mancozeb has been very effective.

Nitrogen toxicity

Figure 301

Cause Excessive fertilizer rate (especially nitrogen)

Signs and symptoms Fronds appear multilobed with indentations. Tips of fronds may be crinkled and clear and sometimes dead. This is primarily a problem on bird's nest fern (*Asplenium*) and closely related species of ferns.

Control Reducing the amount of fertilizer applied and leaching the potting medium with water are recommended for this disorder. In addition, plants may be transplanted to new potting media and left unfertilized until normal growth returns. Sometimes damage is so severe that affected plants must be discarded. Recommended fertilizer rates for bird's nest fern are 1,200 lb of N/acre/year of a 1-1-1 formulation, 2.5 lb of N/1,000 ft^2/month, or 100 to 200 ppm of N in a constant-feed program.

Selected Reference

Poole, R. T., and C. A. Conover. 1983. Fertilization of bird's nest fern. University of Florida, Agricultural Research Center-Apopka, ARC-A Research Report, RH-83-18.

Pseudomonas blights

Figures 302, 303, and 304

Cause *Pseudomonas cichorii* or *P. gladioli*

Signs and symptoms Leaf spots caused by these two bacterial pathogens are nearly identical, and the diseases can be treated in the same manner. Small, water-soaked, translucent spots form on leaves. The spots rapidly enlarge to 2 mm in diameter and turn red brown with a purple margin. When conditions are wet and warm, these areas merge and may spread along veins, encompassing large portions of the fronds. They can be vein delimited and spread over one side of a leaf without crossing the central leaf vein.

Control Elimination of overhead watering is one of the most effective control methods for bacterial blights. Blight caused by *Pseudomonas gladioli* is more aggressive than that caused by *P. cichorii* and is consequently more difficult to control, even when leaves are kept dry. Applications of bactericides such as copper or antibiotic products are generally ineffective. Always use pathogen-free plants for production, and destroy symptomatic plants as soon as they are discovered.

Selected References

Ark, P. A., and C. M. Tompkins. 1946. Bacterial leaf blight of bird's-nest fern. Phytopathology 36:758-761.

Chase, A. R., J. W. Miller, and J. B. Jones. 1984. Leaf spot and blight of *Asplenium nidus* caused by *Pseudomonas gladioli*. Plant Dis. 68:344-347.

Rhizoctonia aerial blight and leaf spot

Figures 305, 306, and 307

Cause *Rhizoctonia solani*

Signs and symptoms Rhizoctonia aerial blight occurs primarily during the summer or warmer months. Disease development can occur in less than a week, so plants should be checked carefully and frequently. Brown, irregularly shaped spots form anywhere on the foliage but most commonly within the crown of the plant, which is often wet. Sometimes the first symptoms form near the top of plant, confusing the source of the disease (the soil). The disease spreads rapidly, and the entire plant can become covered with the brown, weblike mycelium of the pathogen.

Control A pathogen-free potting medium is the first step in the control of all soilborne pathogens such as *Pythium* and *Rhizoctonia*. Plants should be produced from pathogen-free stock and grown in new or sterilized pots on raised benches. Since this pathogen inhabits the soil, both the roots and the foliage of the plants must be treated with a fungicide to provide optimal disease control. Chlorothalonil, iprodione, and thiophanate methyl have provided excellent control of Rhizoctonia aerial blight on Boston fern experimentally. Air temperatures above 95°F (35°C) and soil temperatures above 90°F (32°C) curtail disease development. Although many other foliage plants are hosts of *Rhizoctonia* spp., an isolate from one host may or may not infect another host. It is important, therefore, to protect all foliage plants from this disease should it be found on any crop.

Selected References

Chase, A. R. 1991. Characterization of *Rhizoctonia* species isolated from ornamentals in Florida. Plant Dis. 75:234-238.

Chase, A. R., and C. A. Conover. 1986. Temperature and potting medium composition affect Rhizoctonia aerial blight of Boston fern. University of Florida, Agricultural Research and Education Center-Apopka, AREC-A Research Report, RH-86-6.

Chase, A. R., and C. A. Conover. 1987. Temperature and potting medium effects on growth of Boston fern infected with *Rhizoctonia solani*. HortScience 22:65-67.

Knauss, J. F. 1971. Rhizoctonia blight of 'Florida Ruffle' fern and its control. Plant Dis. Rep. 55:614-616.

Polyscias and Related Plants

Dizygotheca

False aralias are native to the islands of New Caledonia and Polynesia, where many attain the stature of trees. As foliage plants, they may be used as small, desk-top plants or as large, multitrunked specimen plants. Plants are produced with 2,000 to 4,000 ft-c and a minimum temperature of 65°F (18°C). False aralia can be maintained indoors with a minimum of 150 ft-c. Diseases of false aralia are not common, but mites, mealybugs, and scales can be problems.

Fatshedera

Fatshedera lizei (the sole member of this genus) was created by crossing *Fatsia japonica* and *Hedera helix*. Fatshederas are used as low floor plants and are especially tolerant of low temperatures. They are propagated from stem cuttings and grown at light levels of 4,000 to 6,000 ft-c. Greenhouse night temperatures should be maintained above 50°F (10°C) to allow growth of this plant. Indoor light levels of 100 to 150 ft-c are recommended. These plants are relatively resistant to most diseases but are hosts of both spider and tarsonemid mites, mealybugs, aphids, and scales.

Fatsia

Japanese aralia is a large shrub native to Japan. Indoors, they are used primarily as specimen plants. Plants are usually produced from seeds under approximately 4,000 to 6,000 ft-c, with a minimum temperature of 50°F (10°C). Indoor light levels of 150 ft-c are needed for maintenance of plant quality. Fatsias are hosts of several bacterial and fungal pathogens that affect leaves, stems, and roots. Pests of *Fatsia* include mealybugs, both spider and tarsonemid mites, aphids, and scales.

Polyscias

Aralias range from small trees to shrubs in their native environment of the South Sea islands and tropical Asia. Most aralias are used as floor or specimen plants, although some grow best as small, desk-top plants. Plants are most commonly produced with 1,500 to 4,500 ft-c and temperatures between 70 and 85°F (21 and 30°C). Aralias require at least 150 ft-c in the interiorscape. These plants are subject to many of the same diseases as their close relatives, the scheffleras, although they are generally less susceptible. Mites are the most serious pest of aralias, although mealybugs, scales, and aphids can also be found on them.

Alternaria leaf spot

Figures 308 and 309

Cause *Alternaria panax*

Signs and symptoms Alternaria leaf spot is known to occur on *Dizygotheca, Fatsia, Fatshedera, Polyscias,* and

Tupidanthus spp. Lesions are initially tiny, translucent areas that later turn tan and have a chlorotic halo. Sometimes a reddish purple pigment is associated with leaf spot of variegated *Polyscias balfouriana*. Lesions on *P. fruticosa* can enlarge to 1 cm and cause severe defoliation within 5 days of infection. Large, dark brown to black leaf spots appear anywhere on the leaf and sometimes on petioles and stems. Severe infections commonly result in leaf drop and give the plant a sparse appearance that can be confused with a similar leaf loss associated with root rot.

Control   Keeping leaves dry will completely control this disease without any need for fungicides. Disease occurs at temperatures between 65 and 86°F (18 and 30°C) but is most severe between 75 and 80°F (21 and 27°C). Alternaria leaf spot can be reduced on scheffleras by applying fertilizer at rates higher than recommended, but similar tests showed that Alternaria leaf spot on false aralia (*Dizygotheca*) was not affected by fertilizer rate and that higher rates severely reduced plant quality.

Selected References

Atilano, R. A. 1983. A foliar blight of Ming aralia caused by *Alternaria panax*. Plant Dis. 67:224-226.

Atilano, R. A. 1983. Increase of Alternaria blight in two ornamental foliage plant species treated with benomyl. Plant Dis. 67:804-805.

Chase, A. R. 1984. Alternaria leaf spot control with chemicals. Foliage Digest 7(11):3.

Chase, A. R. 1986. Efficacy of Chipco 26019 flowable and wettable powder formulations for control of Alternaria and Drechslera leaf spots. Univ. of Florida, Agricultural Research and Education Center-Apopka, AREC-Apopka Research Report, RH-86-2.

Chase, A. R., and R. T. Poole. 1986. Effects of fertilizer rate on severity of Alternaria leaf spot of three plants in the Araliaceae. Plant Dis. 70:1144-1145.

Uchida, J. Y., M. Aragaki, and M. A. Yoshimura. 1984. Alternaria leaf spots of *Brassaia actinophylla*, *Dizygotheca elegantissima*, and *Tupidanthus calyptratus*. Plant Dis. 68:447-449.

Anthracnose

Figure 310

Cause   *Colletotrichum gloeosporioides*

Signs and symptoms   Lesions are initially water soaked and surrounded by a yellow halo. Eventually, they can reach 3 cm in diameter and turn tan to black. The tiny, black fruiting bodies of the pathogen are readily detected in the lesions on the upper leaf surface. Most lesions occur along leaf margins or in wounded areas.

Control   Elimination of overhead watering and exposure to rainfall can reduce disease incidence and severity. Mancozeb alone or in combination with thiophanate methyl can be very effective in controlling this disease. Check fungicide labels for specific crops, application rates, and intervals.

Selected Reference

Aragaki, M., K. Y. Pitz, and J. Y. Uchida. 1985. Foliar blight and damping-off of *Brassaia actinophylla* caused by *Colletotrichum gloeosporioides*. (Abstr.) Phytopathology 75:1382.

Crown gall

Figure 311

Cause   *Agrobacterium tumefaciens*

Signs and symptoms   Slightly swollen areas on the stems, leaf veins, and even roots are initially apparent. These swollen areas enlarge and become corky. In cases of severe infection, they may enlarge and merge to create a very distorted stem or root mass. Galls may also form on the ends of cuttings or on stems where cuttings have been removed.

Control   Remove and destroy all plants found infected with the bacterium, and then sterilize any cutting tools used on them. Since a number of fungi can cause galls, an accurate disease diagnosis must be made before choosing control strategies. Never propagate from plants with galls, since the disease will be transferred to each new crop of cuttings.

Fusarium stem rot

Figures 312 and 313

Cause   *Fusarium* spp.

Signs and symptoms   Fusarium stem rot typically appears as a soft, mushy rot at the base of a cutting or rooted plant. The rotten area frequently has a purplish to reddish margin. *Fusarium* sometimes forms tiny, bright red, globular structures (fruiting bodies) at the stem base of a severely infected plant. A wilt disease has been reported from Great Britain but has not been seen in the United States.

Control   Remove infected plants from stock areas as soon as they are detected. Since Fusarium stem rot is similar in appearance to Erwinia blight, accurate disease diagnosis is very important prior to applications of pesticides. If stem rot or cutting rot is a problem, treatment of the cuttings with a dip or a post-sticking drench should diminish losses. Products containing thiophanate methyl provide control of Fusarium stem rot.

Selected Reference

Triolo, E., and G. Lorenzini. 1983. *Fusarium oxysporum* f. sp. *fatshederae*, a new forma specialis causing wilt of × *Fatshedera lizei*. Ann. Appl. Biol. 102:245-250.

Phytophthora aerial blight

Figure 314

Cause   *Phytophthora parasitica*

Signs and symptoms   Lesions are dark brown, water soaked, and irregularly shaped. The disease is most severe during the summer months in shade houses.

Control Growing plants in new or sterilized media on raised benches eliminates much of the source of this disease. Other plants susceptible to this pathogen include dieffenbachia, spathiphyllum, and philodendron.

Pseudomonas leaf spot

Figures 315, 316, and 317

Cause *Pseudomonas cichorii*

Signs and symptoms Spots are found primarily along the leaf margins of dwarf schefflera and are initially small, water-soaked areas that rapidly enlarge and turn black. Severe leaf drop is common on some hosts, and the overall appearance of plants is quite similar to that of plants infected with Alternaria leaf spot. This disease is most common on dwarf schefflera, but *Fatsia, Fatshedera,* and false aralia can also be infected.

Control Bactericides are often used but are not very effective. Control must be based on keeping foliage dry and removing infected leaves or plants from the growing area to reduce spread to healthy plants.

Selected Reference

Chase, A. R., and D. D. Brunk. 1984. Bacterial leaf blight incited by *Pseudomonas cichorii* in *Schefflera arboricola* and some related plants. Plant Dis. 68:73-74.

Root-knot nematode

Figure 318

Cause *Meloidogyne* spp.

Signs and symptoms Galls occur on roots, and the root system may be drastically reduced. Plant stunting and wilting occur when infestations are severe.

Control Use sterile soil, and grow plants off the ground if possible. Check the roots of all plants and rooted cuttings for root-knot nematode infestation before bringing them into your nursery or landscape. Some nematicides are safe and effective for root-knot nematode control; check the labels for your plant and application methods.

Xanthomonas leaf spot

Figures 319, 320, and 321

Cause *Xanthomonas campestris* pv. *hederae*

Signs and symptoms Xanthomonas leaf spot of aralia has become more common during the past 5 years. Lesions are initially tiny, corky-appearing areas on the lower leaf surface that can be easily mistaken for symptoms of edema or phytotoxicity. The lesions generally enlarge to 2 to 15 mm in diameter and become black with a chlorotic margin. Infections of *Fatsia japonica* are generally confined to pinpoint, yellow to tan lesions scattered across the leaf surface. Lesions are rarely more than 1 mm wide and have irregularly shaped, raised edges, giving the lower leaf surface a corky appearance. Severe infections of lower leaves often result in complete chlorosis and leaf abscission. Infections of immature leaves of Japanese aralia result in severe deformity, which can resemble damage from insect feeding.

Control The following plants are known hosts of this bacterium: *Dizygotheca elegantissima* (false aralia), dwarf schefflera, English ivy, Japanese aralia, *Polyscias fruticosa* (Ming aralia), and schefflera. Chemical control with bactericides is rarely effective. Minimizing overhead irrigation and using pathogen-free stock are keys to control of Xanthomonas leaf spot on aralias and their relatives.

Selected Reference

Chase, A. R. 1984. *Xanthomonas campestris* pv. *hederae* causes a leaf spot of five species of Araliaceae. Plant Path. 33:439-440.

Radermachera

China doll (radar plant) is an evergreen tree native to southeast Asia. Plants grow best under 2,000 to 3,000 ft-c. They are hosts of a number of fungal stem diseases as well as broad mites, mealybugs, caterpillars, and aphids.

Corynespora leaf spot

Figure 322

Cause *Corynespora cassiicola*

Signs and symptoms Lesions expand rapidly, are black, and may encompass the entire leaflet and cause abscission when conditions are favorable. There is rarely any halo surrounding lesions on China doll.

Control Keep plants as free of excess water as possible, and avoid crowding to promote rapid drying of foliage. Mixed results in controlling this disease have been seen with the fungicides available for Corynespora leaf spot on other plants.

Damping-off

Figure 323

Cause *Fusarium, Rhizoctonia, Pythium,* and *Phytophthora* spp.

Signs and symptoms Poor germination, blackening of roots, or mushiness can be followed by yellowing, wilting, and loss of the plant or seedling. An overall poor stand results, which makes repotting or discarding necessary.

Control These organisms usually have wide host ranges and frequently affect many plant species in the same nursery. Always use clean seed and new potting media and containers, and whenever possible, grow plants on raised benches to limit exposure to native pathogens in the soil under the pots. The soil moisture should be maintained as low as possible to reduce pathogen growth without reducing plant growth. Preventive drenches with a variety of fungicides are often chosen by growers as extra insurance against damping-off. Be sure to obtain an accurate disease diagnosis, since fungicides that are effective against *Fusarium* and *Rhizoctonia* spp. differ from those for *Pythium* and *Phytophthora* spp.

Fusarium stem rot and blight

Figure 324

Cause *Fusarium* spp.

Signs and symptoms Fusarium stem rot typically appears as a soft, mushy rot at the base of a cutting or rooted plant. The rotten area often has a purplish to reddish margin. *Fusarium* sometimes forms tiny, bright red, globular structures (fruiting bodies) at the stem base of a severely infected plant.

Control Use pathogen-free seeds and new or sterilized pots and potting media, and grow plants away from the native soil on raised beds or benches, if possible. Remove infected plants from stock areas as soon as they are detected. It is necessary to obtain a laboratory diagnosis for stem rot of China doll, since several other fungi can cause similar symptoms and optimal chemical controls are very different depending upon the cause of the disease.

Rhizoctonia stem rot and blight

Figure 325

Cause *Rhizoctonia solani*

Signs and symptoms Rhizoctonia stem rot is typified by lesions at the soil line on stems of small seedlings. Lesions are tan to black, dry, and shrunken. When lesions encompass the entire stem, the seedling wilts and dies. Larger plants rarely seem affected by this disease and apparently can outgrow an early infection.

Control Use the control methods listed for damping-off.

Selected References

Chase, A. R. 1991. Characterization of *Rhizoctonia* species isolated from ornamentals in Florida. Plant Dis. 75:234-238.
Chase, A. R. 1992. Efficacy of thiophanate methyl fungicides for diseases of Florida ornamentals. Proc. Fla. State Hortic. Soc. 105:182-186.
Chase, A. R., and T. A. Mellich. 1992. Controlling *Rhizoctonia* diseases on ornamentals with fungicides. University of Florida, Central Florida Research and Education Center-Apopka, CFREC-A Research Report, RH-92-8.

Rhoeo

Rhoeos are native to the moist regions of the West Indies, Guatemala, and Mexico. They are produced under 4,000 to 6,000 ft-c and should not be exposed to temperatures below 55°F (13°C). Rhoeos are subject to a few diseases and, more commonly, to mealybugs and spider mites.

Curvularia leaf spot

Figure 326

Cause *Curvularia eragrostidis*

Signs and symptoms Spots are initially small, green, and sunken and form primarily on the lower leaf surface. They enlarge to 7 mm or larger and turn tan (thus the common name, tan leaf spot). When leaves are infected before fully expanding, the fungus can cause distortion of the leaves.

Control Eliminate overhead irrigation and exposure to rainfall if possible. Fungicide sprays with chlorothalonil, thiophanate methyl, or mancozeb were effective in controlling tan leaf spot experimentally.

Selected Reference

Miller, J. W. 1971. Tan leaf spot of *Rhoeo discolor* caused by *Curvularia eragrostidis*. Plant Dis. Rep. 55:38-40.

Tobacco mosaic

Figure 327

Cause Tobacco mosaic virus (TMV)

Signs and symptoms Typical mosaic symptoms are light and dark green areas on infected leaves. Leaves are sometimes distorted, and stunting or poor growth may occur in severe cases.

Control Discard all plants with TMV symptoms. They must not be used as stock for propagation, since all cuttings removed from them will have the disease as well.

Selected Reference

Thompson, S. M., and M. K. Corbett. 1985. Mosaic disease of *Rhoeo discolor* caused by a strain of tobacco mosaic virus. Plant Dis. 69:356-359.

Saintpaulia (African violet)

African violets are frequently grown by producers of foliage plants under approximately the same conditions as those used for many other indoor crops. They require 1,000 ft-c for good growth and continued blooming indoors; a minimum temperature of 70°F (21°C) is recommended. Plants should be irrigated with water at the same temperature as their leaves, since applications of cold water to warmer leaves result in white rings. African violets are hosts of many viral, bacterial, fungal, and nematode diseases that affect their leaves, flowers, stems, and roots. Pests include foliar nematodes, mealybugs, scales, and tarsonemid mites.

Botrytis blight

Figures 328 and 329

Cause *Botrytis cinerea*

Signs and symptoms Lesions usually appear on the leaf underside, especially on petioles near the pot rim or in contact with the potting medium. A small, water-soaked lesion can rapidly enlarge and extend into the blade or petiole, causing its collapse. Sporulation on necrotic leaf or flower tissue is characterized by a powdery, grayish brown mass of conidia.

Control Avoid growing conditions characterized by low light, high humidity, poor air circulation, and warm days with cool nights. Fungicides such as vinclozolin effectively control Botrytis blight. Although iprodione is effective against Botrytis blight, it may cause stunting and chlorosis on some African violet cultivars.

Selected Reference

Beck, G. E., and J. R. Vaughan. 1949. Botrytis leaf and blossom blight of *Saintpaulia*. Phytopathology 39:1054-1056.

Chimera

Figure 330

Cause Genetic variability in the plant

Signs and symptoms Leaves develop yellow variegation that can be mistaken for symptoms of viral infections. Plants with these symptoms usually occur very rarely within a group in contrast to those infected with a virus, which may be more numerous. Distortion can also occur, especially on some cultivars produced in tissue culture where the plant genetics may become abnormal because of the unusual growing conditions.

Control If the number of off-type plants is high, a new source of plants should be found. Discard those that are found, but do not miss the opportunity to develop a new selection of the plant. This is one of the oldest ways for new plants to come into the commercial trade.

Cold water damage

Figure 331

Cause Extremes in water temperature

Signs and symptoms White, light yellow, or pale green round spots (sometimes like donuts) appear on the upper leaf surface. Spots can appear along margins or in the centers and are sometimes irregular or donut-shaped and white.

Control Water that is colder than the leaf surface will cause spotting; this injury is most common during the winter. Be sure the temperature of the water applied from overhead is near the leaf-surface temperature, or water plants via capillary mats or another system that avoids leaf wetting.

Selected References

Elliott, F. H. 1946. *Saintpaulia* leaf spot and temperature differential. Proc. Am. Soc. Hortic. Sci. 47:511-514.

Poesch, G. H. 1942. Ring spot on *Saintpaulia*. Proc. Am. Soc. Hortic. Sci. 41:381-382.

Corynespora leaf spot

Figure 332

Cause *Corynespora cassiicola*

Signs and symptoms Lesions appear first as tiny, sunken, slightly brown areas. These areas can enlarge to 2.5 cm in diameter and darken with age. This disease is found on other gesneriads such as *Nematanthus* and *Columnea* spp. and *Aeschynanthus pulcher* as well as on some cultivars of *Ficus benjamina*.

Control Use the same cultural controls as mentioned for Botrytis blight. Chemical control trials have indicated that both mancozeb and chlorothalonil provide excellent disease control on some hosts.

Selected Reference

Chase, A. R. 1982. Corynespora leaf spot of *Aeschynanthus pulcher* and related plants. Plant Dis. 66:739-740.

Erwinia blight

Figure 333

Cause *Erwinia chrysanthemi*

Signs and symptoms *Erwinia chrysanthemi* can infect all tissues of African violet. Root infection is characterized by a rotted, water-soaked root system. A crown rot is also found sometimes. Infected petioles and leaves are greasy brown to black. Wilting and complete collapse of plants are symptoms of advanced disease.

Control Although there is a wide range of susceptibility levels to bacterial blight in African violet cultivars, most are moderately susceptible. Since bacterial diseases are difficult to control with the bactericides currently available, disease avoidance is very important. Minimize irrigations to reduce water splashing of bacteria, and discard all plant materials with suspected bacterial infections. This pathogen infects a wide range of other foliage plants.

Selected References

Knauss, J. F., and J. W. Miller. 1974. Bacterial blight of *Saintpaulia ionantha* caused by *Erwinia chrysanthemi*. Phytopathology 64: 1046-1047.

Roberts, B. J. 1977. Susceptibility of certain *Saintpaulia* species and cultivars to bacterial blight. Plant Dis. Rep. 61:1048-1050.

Ethylene damage

Figures 334 and 335

Cause Exposure to ethylene gas

Signs and symptoms Plants develop watery, black spots on leaves and petioles and even in flower centers. The spots look the same as those caused by some fungi and bacteria.

Control These symptoms can appear during shipping (or soon after) when plants are exposed to ethylene. Since fruits and vegetables generate this gas, avoid shipping African violets with produce. Storing plants under high humidity and temperature conditions can also promote ethylene damage.

Selected References

Blanpied, G. D. 1985. Ethylene in postharvest biology and technology of horticultural crops (Symposium). HortScience 20:39-60.

Hanan, J. J. 1973. Ethylene pollution from combustion in greenhouses. HortScience 8:23-24.

Hack, W. W., and E. G. Pires. 1962. Effect of ethylene on horticultural and agronomic plants. MP-613. The Agric. and Mech. Coll. of Texas, Texas Agric. Exp. Stn., College Station, TX.

Foliar nematode

Figure 336

Cause   *Aphelenchoides ritzemabosi*

Signs and symptoms   Small, tan, interveinal sunken areas appear on lower leaf surfaces. These lesions eventually are visible on the upper leaf surfaces as well. Lower leaf surfaces become shiny, brown, and slightly cupped. Severe reduction in leaf size as well as distortion are also common.

Control   It is always best to discard plants suspected of foliar nematode infestations to avoid spreading the nematode to healthy plants by handling. In addition, never use nematode-infested plants as stock, since this will perpetuate the problem. Both preventive and eradicative treatments with either aldicarb or oxamyl have been tested experimentally and were effective and safe for African violets infested with foliar nematodes.

Selected Reference

Strider, D. L. 1979. Control of *Aphelenchoides ritzemabosi* in African violet. Plant Dis. Rep. 63:378-382.

Phytophthora stem and root rot

Figures 337 and 338

Cause   *Phytophthora parasitica*

Signs and symptoms   Symptoms of Phytophthora stem and root rot are very similar in appearance to those of bacterial blight caused by *Erwinia chrysanthemi*. Mixed infections with the two pathogens sometimes occur. Culture of the pathogen is necessary prior to developing a control program for either disease.

Control   Avoid overwatering, since waterlogged roots are easily attacked by *P. parasitica*. As with all diseases caused by soilborne pathogens, use pathogen-free pots, potting media, and plant material. Some differences in cultivar resistance to *P. parasitica* have been found; certain cultivars are very resistant. Etridiazole, metalaxyl, and propamocarb are available for Phytophthora stem and root rot of African violet. Since the pathogen is present in the root system and/or potting medium, these fungicides must be applied to the potting medium to achieve disease control.

Selected References

Krober, H., and H. P. Plate. 1973. Phytophthora-Faule an *Saintpaulien* [Erreger: *Phytophthora nicotianae* var. *parasitica* (Dast.) Waterh.]. Phytopathol. Z. 76:348-355.

Strider, D. L. 1978. Reaction of African violet cultivars to *Phytophthora nicotianae* var. *parasitica*. Plant Dis. Rep. 62:112-114.

Strider, D. L. 1979. Evaluation of Melodie and Optimara African violets for resistance to Phytophthora rot. Plant Dis. Rep. 63:382-384.

Powdery mildew

Figure 339

Cause   *Oidium* sp.

Signs and symptoms   Lesions appear on flowers, petioles, and leaves. A powdery, white coating can form circular areas up to 1 cm in diameter as single lesions or can coalesce to cover the entire leaf.

Control   Cultivars differ in susceptibility to powdery mildew. The disease apparently does not cause serious losses in some states, since many growers do not apply fungicides during an outbreak. Triadimefon is effective for powdery mildew control on African violet.

Selected Reference

Strider, D. L. 1980. Resistance of African violet to powdery mildew and efficacy of fungicides for control of the disease. Plant Dis. 64:188-190.

Pythium root rot

Figure 340

Cause   *Pythium* spp.

Signs and symptoms   Leaves wilt, may turn yellow or pale green, and eventually die. Plants are frequently stunted, and examination of the roots reveals their rotted condition. Initial infections of the roots appear as small, water-soaked, gray or brown areas. These spots can rapidly expand to affect the entire root system. Severely infected plants may have no living roots remaining by the time they are examined.

Control   Prevention is always the best control of a soilborne pathogen like *Pythium*. Use clean pots and potting media, and grow plants on raised benches. Since African violets are rarely tolerant of heavy or poorly draining potting media, the appropriate mix is critical. Even regular fungicide applications to infected plants in a heavy potting medium will not control this disease. Fungicides registered for African violets that should aid in control of this root rot are listed above under Phytophthora stem and root rot.

Root-knot nematode

Figure 341

Cause   *Meloidogyne* spp.

Signs and symptoms Galls occur on roots, and the root system may be drastically reduced; plant stunting and wilting occur when severe infestations are present.

Control Use sterile soil, and grow plants off the ground if possible. Oxamyl will aid in control, but be sure to check labels for this plant and legal application methods before use.

Tomato spotted wilt

Figure 342

Cause Tomato spotted wilt virus (TSWV)

Signs and symptoms TSWV can infect many flowering and foliage potted plants; vegetables such as tomato and lettuce can also be affected. Symptoms on African violet are dramatic; concentric rings of sunken tissue, which can be black, appear. These spots look like waves and are generally easier to see on leaf undersides. This virus is spread by certain thrips, including the western flower thrips, as well as by using infected leaves to propagate new plants.

Control Thrips control must be the first step in controlling this viral disease. Since the host range of TSWV is so large, all new cuttings or stock plants should be examined before they are brought into a greenhouse. Do not use any plants or parts that come from stock with these symptoms.

Selected Reference

Hausbeck, M. K., R. A. Welliver, M. A. Derr, and F. E. Gildow. 1992. Tomato spotted wilt virus survey among greenhouse ornamentals in Pennsylvania. Plant Dis. 76:795-800.

Sansevieria

Snake plants are native to Africa, Arabia, and India. These plants are ideal for indoor use, since they tolerate very low light (50 ft-c), infrequent watering, and low humidities. Interior use can range from dish gardens to large specimen plants. Stock production of snake plants usually occurs under full sun; potted plants are grown under 1,500 to 6,000 ft-c. Good growth occurs over a temperature range of 70 to 90°F (21 to 32°C). Chilling damage occurs when plants are exposed to 32 to 50°F (0 to 10°C). Snake plants are subject to relatively few diseases and pests when produced properly, although cutting rots caused by a variety of fungi can be common. Thrips also can cause losses during production of this plant.

Aspergillus rhizome rot

Figure 343

Cause *Aspergillus niger*

Signs and symptoms Water-soaked spots on rhizomes become sunken as they mature. The epidermis may remain intact while underlying tissue disintegrates rapidly. Lesions

near leaf bases or stem ends form and frequently contain the sooty, dark brown to black spores of the pathogen. The pathogen enters cuttings through wounds created during the process of removing cuttings. The disease can cause severe cutting loss under the warm, moist conditions of shipment.

Control Reduction of disease through maintenance of soil temperatures below 100°F (by using mulches of bark, sawdust, or wood chips) is partially successful. Examine cuttings carefully for signs of infection, and reject those that are not completely healthy. Use new or sterilized pots and potting media, and grow plants away from the native soil.

Selected References

Alvarez Garcia, L. A., and M. A. Diaz. 1949. Aspergillus root-stalk rot of sansevieria (*Sansevieria laurentii*, Wildem). Univ. Puerto Rico J. Agric. 33(1):45-53.

Chase, A. R., and C. A. Conover. 1981. Rhizome rot of *Sansevieria* spp. caused by *Aspergillus niger*. Foliage Digest 4(7):3.

Chilling damage

Figure 344

Cause Temperatures below 45°F (7°C)

Signs and symptoms Leaves are constricted in width in a narrow band across the blade but usually have normal coloration. The leaves develop whitish, water-soaked areas 1 to 4 weeks after their exposure to cold.

Control Maintain the production environment at 45°F (7°C) or above. Prevent plant exposure to cold air or cold water (condensate or drip-through from the roof of the structure). Low air temperatures account for most of the injury in open stock beds, while cold water draining through perforations in polyethylene film used for lining shade houses accounts for most of the injury to sansevieria in central Florida. Use solid-cover structures that are properly heated. Plants that are overfertilized (especially with nitrogen) are more sensitive to chilling than those fertilized appropriately.

Selected References

Conover, C. A., and R. T. Poole. 1976. Influence of nutrition on yield and chilling injury of *Sansevieria*. Proc. Fla. State Hortic. Soc. 89:305-307.

Marlatt, R. B. 1974. Chilling injury in *Sansevieria*. HortScience 9:539-540.

Fusarium leaf spot

Figure 345

Cause *Fusarium moniliforme*

Signs and symptoms Lesions initially appear as tiny, water-soaked areas on immature leaves and rapidly enlarge into elliptical to irregularly shaped areas with red or tan coloration. The lesions are often surrounded by a bright yellow halo, which can be 2 mm wide. As lesions coalesce, immature leaves become distorted, and depending upon the location of the lesions on the leaves, distal portions may become completely necrotic. Under optimal conditions, meristems of infected shoots die.

Control   Standing water in the central whorl of the plant is necessary for infection to occur. The high levels of water available during the rooting process are optimal for disease initiation and progress in cuttings. Complete disease control can be achieved if plant foliage is kept dry. Dracaenas are also susceptible to this pathogen. Chlorothalonil was especially effective in controlling Fusarium leaf spot experimentally. Check labels for legal use on specific plants, intervals, and rates.

Selected References

Chase, A. R. 1981. Update: Fusarium leaf spot of dracaenas and sanseverias. Foliage Digest 4(1):14.

Chase, A. R., and L. S. Osborne. 1981. Pesticide evaluations in support of registrations on tropical foliage plants. University of Florida, Agricultural Research Center-Apopka, ARC-A Research Report, RH-81-14.

Jones, L. K. 1940. Fusarium leaf spot of *Sansevieria*. Phytopathology 30:527-530.

Soft rot

Figure 346

Cause   *Erwinia carotovora*

Signs and symptoms   A mushy soft rot occurs at the lower end of a cutting. Sometimes the plants have a fishy, rotten odor, characteristic of *Erwinia* infections.

Control   Bacterial leaf spot can be controlled through elimination of water on leaves. Choice of clean cuttings and strict sanitation are the most important control measures. Use of bactericides is not recommended because efficacy is very poor. Many other foliage plants can be infected by *Erwinia* spp., and all should be inspected regularly for soft rot symptoms.

Selected Reference

Brown, J. G., and A. M. Boyle. 1944. Bacterial soft rot of sansevieria. Phytopathology 34:350-351.

Schefflera

Scheffleras are native to the tropical and subtropical regions of Asia and the South Sea islands. Many members of this group, such as the umbrella tree (*Brassaia actinophylla*), are trees and shrubs in their native habitat. Scheffleras are used as specimen plants because of their large size and showy leaf shapes. They are produced usually from seeds or occasionally from stem cuttings and tissue-cultured plantlets at light levels of 5,000 to 7,000 ft-c and temperatures between 65 and 90°F (18 and 32°C). Some growers also produce this plant in full sun. Scheffleras

require at least 150 ft-c light in the interior, which restricts their use. Scheffleras are subject to many bacterial and fungal diseases that attack roots, stems, and leaves and many mite, scale, aphid, and mealybug pests. The most serious pests of schefflera in the interiorscape are spider mites. *B. actinophylla* is one of the most pesticide-sensitive foliage plants produced, making pest and disease control difficult.

Acephate phytotoxicity

Figure 347

Cause   Soil application of acephate

Signs and symptoms   Leaves develop marginal necrosis and readily drop from the plant.

Control   Do not apply acephate to the potting medium of scheffleras. Foliar sprays have been safely used, but a foliar spray that causes large amounts of runoff from leaves can also result in these symptoms. Be sure to apply all pesticides as directed by their labels.

Selected Reference

Osborne, L. S., and A. R. Chase. 1984. Influence of acephate and oxamyl on *Alternaria panax* and on Alternaria leaf spot of schefflera. Plant Dis. 68:870-872.

Alternaria leaf spot

Figures 348 and 349

Cause   *Alternaria panax*

Signs and symptoms   Large, dark brown to black leaf spots appear anywhere on the leaf and sometimes on petioles and stems. Severe infections commonly result in leaf drop and give the plant a sparse appearance that can be confused with similar leaf loss associated with root rot. Leaf spots appear wet and can spread in a few days to encompass the entire leaf. Splashing water can carry conidia of *A. panax* from abscised leaves on the ground to leaves several feet above the ground. The pathogen is active between 64 and 86°F (17 and 30°C) but causes the most damage at 75 to 81°F (24 to 27°C).

Control   Keeping the foliage dry will completely control this disease without any need for fungicides. Using fertilizer at rates higher than recommended also significantly reduces disease severity on both schefflera and dwarf schefflera. Since scheffleras are very prone to phytotoxic reactions, characterized by distortion of new leaves, only a few fungicides are both safe and effective on this plant. Many chemicals control Alternaria leaf spot on schefflera but are also somewhat phytotoxic, including chlorothalonil, which is very effective but cannot be used on schefflera or dwarf schefflera because it causes leaf distortions. Mancozeb and iprodione are also very effective and safe to use on schefflera. Although benomyl was shown to significantly increase severity of Alternaria leaf spot of schefflera, thiophanate methyl compounds did not have the same effect, whether applied as a soil drench or a foliar spray.

Selected References

Atilano, R. A. 1983. Alternaria leaf spot of *Schefflera arboricola*. Plant Dis. 67:64-66.

Atilano, R. A. 1983. Increase of Alternaria blight in two ornamental foliage plant species treated with benomyl. Plant Dis. 67:804-805.

Chase, A. R. 1982. Influence of irrigation method on severity of selected fungal leaf spots of foliage plants. Plant Dis. 66:673-674.

Chase, A. R. 1984. Alternaria leaf spot control with chemicals. Foliage Digest 7(11):3.

Chase, A. R., and R. T. Poole. 1986. Effects of fertilizer rate on severity of Alternaria leaf spot of three plants in the Araliaceae. Plant Dis. 70:1144-1145.

Uchida, J. Y., M. Aragaki, and M. A. Yoshimura. 1984. Alternaria leaf spots of *Brassaia actinophylla*, *Dizygotheca elegantissima*, and *Tupidanthus calyptratus*. Plant Dis. 68:447-449.

Ammonium toxicity

Figure 350

Cause   Use of fertilizer with a high percentage of ammonium

Signs and symptoms   Leaves develop marginal and interveinal yellowing and browning. They also become cupped and turn downward.

Control   Use fertilizers with a higher percentage of nitrate than ammonium, especially during the winter or at any times when plants are growing slowly. Young seedlings (in less than 15-cm pots) are most sensitive to ammonium toxicity.

Bendiocarb phytotoxicity

Figure 351

Cause   Application of bendiocarb to the potting medium

Signs and symptoms   A single application of bendiocarb to the potting medium results in necrotic spots, tip necrosis, leaf cupping, and distortion.

Control   Always follow pesticide labels for applications rates, intervals, and especially sites. Products meant to be applied to leaves are not always safe when applied to the roots via the potting medium.

Selected Reference

Osborne, L. S., and A. R. Chase. 1986. Phytotoxicity evaluations of Dycarb on selected foliage plants. University of Florida, Agricultural Research and Education Center-Apopka, AREC-A Research Report, RH-86-12.

Chemical phytotoxicity

Figure 352

Cause   Application of a variety of pesticides

Signs and symptoms   Single rings of brown callus or scar tissue that are roughly circular form mainly on leaf undersides. Several insecticides and fungicides have been shown to cause these symptoms.

Control   Always test small ba· es of plants under your growing conditions when trying a new pesticide. Even those labeled for schefflera may be phytotoxic under some conditions. Experiments indicate that plants growing rapidly under optimal temperatures are most sensitive to pesticide application.

Selected References

Chase, A. R., and G. W. Simone. 1985. Phytotoxicity on foliage ornamentals caused by bactericides and fungicides. University of Florida. Plant Pathol. Fact PP-30.

Smith, J. H., and C. Morin. 1978. Chemical injury of schefflera. Calif. Dept. of Food & Agric. Plant Path. Repts. 2:55.

Dodder

Figure 353

Cause   *Cuscuta* spp.

Signs and symptoms   Plants have straw-colored filaments growing randomly over their surfaces. The strands of dodder (a parasitic plant) penetrate the stems and develop knots of tissue and sometimes tiny, yellow flowers.

Control   Dodder is not common in greenhouse operations but could be introduced by collecting schefflera seed from a wild tree that is infested with dodder. Remove and destroy all plants from the nursery (and surrounding areas) with dodder. Chemical control is not appropriate.

Edema

Figure 354

Cause   Imbalance between water use and water uptake

Signs and symptoms   Plants develop warty, raised structures on the outer margins of leaves that are easiest to see on leaf undersides. In advanced stages, the outer 1 cm of the leaf edge becomes light green and is greasy appearing.

Control   Manage water carefully during the cooler times of the year when the plants may not be growing rapidly. Cold soils and warm air temperatures are conducive to development of edema on some potted ornamentals. Bottom heating may help in a greenhouse that has a history of producing plants with edema during the winter months.

Frost damage

Figure 355

Cause   Exposure to frost

Signs and symptoms   Leaves turn a reddish color and cup upward.

Control Protect scheffleras and all tropical plants from frost. Leaves with frost damage will not recover, although new leaves may develop normally.

Mite damage

Figure 356

Cause Spider mites feeding in the interiorscape

Signs and symptoms Patches of dull green to gray tissue develop on schefflera, mainly in the interiorscape. In advanced infestations, the leaflets drop. The leaf undersides show the spider mites and their eggs. The normal speckling that accompanies mite feeding on most plants may not be present.

Control Be sure to accurately diagnose problems before beginning treatment. Remember to turn leaves over when scouting. Follow miticide labels for appropriate use in the greenhouse, nursery, or interiorscape.

Phytophthora leaf spot

Figure 357

Cause *Phytophthora parasitica*

Signs and symptoms Symptoms of Phytophthora leaf spot are essentially the same as those of Alternaria leaf spot, except that the spots generally appear first on lower leaves close to the ground. The disease occurs on scheffleras.

Control Grow plants away from the native soil on raised benches. Use pathogen-free pots and potting media, and minimize exposure to rainfall and overhead irrigation. Many fungicides can be phytotoxic to schefflera, but fosetyl aluminum, etridiazole, and metalaxyl are safe and effective for Phytophthora leaf spot. Treatments must include drenching the potting media and the soil around the pots if the plants are grown on the ground.

Selected Reference

Wisler, G. C., W. H. Ridings, and R. S. Cox. 1978. Phytophthora leaf spot of *Brassaia actinophylla*. Proc. Fla. State Hortic. Soc. 91:240-242.

Pseudomonas leaf spot

Figure 358

Cause *Pseudomonas cichorii*

Signs and symptoms Spots are found primarily on the leaf margins of dwarf schefflera and are initially small, water-soaked areas that rapidly enlarge and turn black. Severe leaf drop is common, and the overall appearance of plants is quite similar to those infected with Alternaria leaf spot. Scheffleras and aralias are also slightly susceptible to this bacterial disease.

Control Bactericides can be used but are not very effective. Control must be based on maintaining dry foliage and removing infected leaves or plants from the growing area to reduce spread to healthy plants. Many other plants, including ficus, ferns, and dieffenbachias, are also hosts of *P. cichorii*. Using fertilizer at rates higher than recommended reduces severity of Pseudomonas leaf spot on dwarf schefflera.

Selected References

Chase, A. R., and D. D. Brunk. 1984. Bacterial leaf blight incited by *Pseudomonas cichorii* in *Schefflera arboricola* and some related plants. Plant Dis. 68:73-74.

Chase, A. R., and J. B. Jones. 1986. Effects of host nutrition, leaf age, and preinoculation light levels on severity of leaf spot of dwarf schefflera caused by *Pseudomonas cichorii*. Plant Dis. 70:561-563.

Pythium stem and root rot

Figure 359

Cause *Pythium splendens* and *P. irregulare*

Signs and symptoms Poor germination and emergence or seedling loss after emergence are common symptoms of damping-off. Many organisms, including *Alternaria, Fusarium, Phytophthora, Pythium* and *Rhizoctonia* spp., can be responsible for this disease, although *Pythium splendens* is by far the most common pathogen of schefflera seeds and seedlings. In older plants, stem rot and root rot are significant problems. Affected root systems are sparse and dark, and they disintegrate when handled because their outer layer separates easily from the inner core.

Control Diagnosis of the causal organism is the most important step toward control of damping-off diseases. Use the same controls as mentioned for Phytophthora leaf spot. Fertilizer at rates higher than recommended reduced severity of Pythium root rot on dwarf schefflera. Research has shown that the best overall fungicidal control is achieved with metalaxyl.

Selected References

Chase, A. R. 1988. Effect of liquid fertilizer rate on severity of Pythium root rot, Alternaria leaf spot and growth of dwarf schefflera, 1987. Biol. and Cult. Tests 3:84.

Chase, A. R., and T. A. Mellich. 1992. Fungicide tests for control of *Phytophthora* and *Pythium* diseases on ornamentals. University of Florida, Central Florida Research and Education Center-Apopka, CFREC-A Research Report, RH-92-9.

Keim, R., T. Mock, R. M. Endo, and E. M. Krausman. 1978. Pathogenicity, characteristics, and host range of the fungus, *Pythium irregulare* Buis. in a container nursery in Southern California. HortScience 13:295-296.

Raabe, R. D., J. H. Hurlimann, and D. S. Farnham. 1978. Root rot control in *Brassaia actinophylla*. Calif. Plant Pathol. 44:2-3.

Yamamoto, B. T., and M. Aragaki. 1983. Etiology and control of seedling blight of *Brassaia actinophylla* caused by *Pythium splendens* in Hawaii. Plant Dis. 67:396-399.

Rhizoctonia blight

Figure 360

Cause   *Rhizoctonia solani*

Signs and symptoms   Rhizoctonia blight occurs primarily during the summer or warmer months. Damping-off can be a serious problem with schefflera seedlings. The disease can develop in less than a week, so plants should be checked carefully and frequently. Brown, irregularly shaped spots form anywhere in the foliage but most commonly within the crown of the plant, which is often wet. The disease spreads rapidly, and the entire plant can become covered with the brown, weblike mycelium of the pathogen.

Control   A pathogen-free potting medium is the first step in the control of all soilborne pathogens. Plants should be produced from pathogen-free stock and grown in new or sterilized pots on raised benches. Since this pathogen inhabits the soil, both the roots and the foliage of the plants must be treated with a fungicide to provide optimal disease control. A combination drench-spray of either thiophanate methyl or iprodione will best accomplish this.

Selected Reference

Chase, A. R. 1991. Characterization of *Rhizoctonia* species isolated from ornamentals in Florida. Plant Dis. 75:234-238.

Xanthomonas leaf spot

Figures 361 and 362

Cause   *Xanthomonas campestris* pv. *hederae*

Signs and symptoms   Symptoms on scheffleras and dwarf scheffleras are generally confined to pinpoint, yellow to tan lesions scattered across the leaf surface, although they can become large and confined between leaf veins. Lesions are mostly 1 mm wide with irregularly raised edges, giving the lower leaf surface a corky appearance. Severe infections of lower leaves often result in complete chlorosis and finally leaf abscission.

Control   Control of this disease can be obtained by minimizing free water on leaves. Examine plants frequently for symptoms, and discard all with Xanthomonas blight. English ivy and aralias are also hosts of the pathogen. Nutritional studies with schefflera and dwarf schefflera have shown that applications of fertilizer at rates higher than recommended produce plants with greater resistance to this disease. In addition, use of a fertilizer source with a 50-50 balance between nitrate and ammonium resulted in higher disease levels than use of sources containing either nitrate or ammonium alone.

Selected References

Blake, J. H., and A. R. Chase. 1988. Effect of ammonium-nitrate ratio on growth and quality of *Brassaia actinophylla* and susceptibility to *Xanthomonas campestris* pv. *hederae*. Proc. Fla. State Hortic. Soc. 101:337-339.

Chase, A. R. 1984. *Xanthomonas campestris* pv. *hederae* causes a leaf spot of five species of Araliaceae. Plant Path. 33:439-440.

Chase, A. R. 1988. Effect of nitrogen and potassium rate on severity of Xanthomonas leaf spot of schefflera, 1987. Biol. and Cult. Tests 3:83.

Chase, A. R., and R. T. Poole. 1987. Effects of fertilizer rates on severity of Xanthomonas leaf spot of schefflera and dwarf schefflera. Plant Dis. 71:527-529.

Spathiphyllum

Spathiphyllums, sometimes called peace lilies, are tropical perennial plants. *Spathiphyllum* spp. are grown in various sizes, ranging from dish gardens to large specimens to mass plantings in malls and office buildings. Spathiphyllums are produced with light levels of 1,500 to 2,500 ft-c from seeds or tissue-cultured plantlets. Good growth occurs at temperatures between 65 and 90°F (18 and 32°C). Spathiphyllums are good indoor plants because they tolerate light intensities of 75 to 100 ft-c. Compared with the other aroids, few diseases occur on this plant. The most important diseases are caused by fungi. The most common pest is the mealybug.

Algae

Figure 363

Cause   Algae growth on leaf surfaces

Signs and symptoms   A light or bright green coating develops, especially on lower leaves or in cupped leaves. The coating can be easily scraped away without damaging the leaf surface beneath.

Control   Minimize overhead irrigation, especially when using a foliar feed system to fertilize. Increasing both air circulation (with fans) and light levels can reduce algal growth on leaves. These algae do not harm the plant directly but can reduce photosynthesis by blocking the light. Copper may aid in controlling algae growth, but effective levels may be phytotoxic to the plant.

Benomyl toxicity

Figure 364

Cause   Use of excessive rates of benomyl

Signs and symptoms   Marginal chlorosis and necrosis (tan to gray) occur.

Control   Using labeled rates has proven safe on this plant. Avoid using benomyl at rates of 12 oz (a.i.)/100 gal or above. As with all pesticides, the label must be followed exactly.

Selected Reference

Chase, A. R., and R. T. Poole. 1988. Cylindrocladium root and petiole rot of *Spathiphyllum* spp. University of Florida, Coop. Ext. IFAS, Bulletin 860. 21 pp.

Chimera

Figure 365

Cause   Genetic variability in the plant

Signs and symptoms   Leaves develop yellow variegation that can be mistaken for symptoms of a viral infection. The number of plants with these symptoms is usually lower than the proportion of a crop that would be symptomatic if it were infected with a virus. Distortion can also occur, especially on some cultivars produced in tissue culture where the plant genetics may become abnormal because of the unusual growing conditions.

Control   If the number of abnormal plants is high, a new source of plants should be found. Discard those that are found, but do not miss the opportunity to develop a new selection of the plant. This is one of the oldest ways for new plants to come into the commercial trade.

Chlorosis

Figure 366

Cause   Low fertilizer rates

Signs and symptoms   Plants develop small leaves that are light green or yellow. Eventually they cease to grow. Root systems are generally well developed and greater than would be expected given the degree of top growth.

Control   Apply fertilizer regularly at recommended rates. Be sure to test the soluble salts of the potting medium before applying fertilizer to plants that are yellow and/or stunted. There are a number of soilborne fungi and nematodes that can cause these symptoms. In addition, on some plants, symptoms of overfertilization can appear similar to those of underfertilization.

Selected Reference

Henny, R. J., A. R. Chase, and L. S. Osborne. 1991. *Spathiphyllum.* University of Florida, Central Florida Research and Education Center-Apopka, CFREC-A Foliage Plant Research Note, RH-91-32.

Cylindrocladium root and petiole rot

Figures 367, 368, 369, and 370

Cause   *Cylindrocladium spathiphylli*

Signs and symptoms   One of the first symptoms of this root rot disease is yellowing of lower leaves, sometimes accompanied by slight wilting. Elliptical, dark brown spots may be found on leaves and petioles; lower portions of the petioles frequently rot. At this stage, plant roots are severely rotted and few healthy roots are found. The tops of such plants are easily removed from the pot without any adhering roots.

Control   Control of this root rot disease must be based first upon use of pathogen-free plants from either tissue

TABLE 14. Susceptibility of some *Spathiphyllum* species and cultivars to *Cylindrocladium spathiphylli*

Species and/or cultivar	Disease severity[a]
floribundum	1.0
floribudum 'Mini'	1.0
cannifolium	1.3
lechlerianum	3.4
wallisii	4.5
Tasson	3.2
Bennett	4.2
Mauna Loa	4.4
Queen Amazonica	3.9

[a] 1 = no symptoms; 2 = 1 to 25% diseased; 3 = 26 to 50% diseased; 4 = 51 to 75% diseased; and 5 = 76 to 100% diseased, usually dead.

culture or seed sources. Using sterilized potting medium and pots and growing plants on clean or disinfested raised benches are also important in reducing the chances of disease development and spread. The *Spathiphyllum* cultivars tested to date have been very susceptible to Cylindrocladium root rot, with the exception of *S. floribundum* (Table 14). This species is a host of the pathogen but is highly resistant and shows little root loss when infected with *C. spathiphylli.* Chemical treatments are not completely effective unless disease pressure is low. Under conditions of high disease pressure (especially prevalent during the summer), triflumizole provides better control than alternatives currently available.

Selected References

Chase, A. R. 1982. Control of Cylindrocladium root rot of *Spathiphyllum.* Proc. Fla. State Hortic. Soc. 95:139-141.

Chase, A. R. 1985. Cylindrocladium root and petiole rot control update. Proc. Fla. State Hortic. Soc. 98:115-118.

Chase, A. R., and R. T. Poole. 1988. Cylindrocladium root and petiole rot of *Spathiphyllum* spp. University of Florida, Coop. Ext. IFAS, Bulletin 860. 21 pp.

Henny, R. J., and A. R. Chase. 1986. Screening *Spathiphyllum* species and cultivars for resistance to *Cylindrocladium spathiphylli.* HortScience 21:515-516.

Schoulties, C. L., and N. E. El-Gholl. 1980. Pathogenicity of *Cylindrocladium floridanum* on *Spathiphyllum* sp. cv. Clevelandii. Proc. Fla. State Hortic. Soc. 93:183-186.

Schoulties, C. L., and N. E. El-Gholl. 1983. Host range and pathogen specificity studies of *Cylindrocladium spathiphylli.* Proc. Fla. State Hortic. Soc. 96:282-284.

Dasheen Mosaic

Figure 371

Cause   Dasheen mosaic virus (DMV)

Signs and symptoms   A mosaic pattern of light green to yellow is found on the new leaves of infected plants at various times during the year. Normally, leaf distortion does not occur. This disease does not appear to cause economically significant losses in spathiphyllum production for the majority of growers.

Control   Other aroids (dieffenbachia, anthurium, and philodendron) are susceptible to DMV and can be sources of infection for spathiphyllum. Always discard and destroy plants with these symptoms, and control aphids, since they serve as vectors of the virus. Sometimes symptoms of genetic disorders are similar to those of DMV.

Erwinia blight

Figures 372 and 373

Cause   *Erwinia chrysanthemi* and/or *E. carotovora* subsp. *carotovora*

Signs and symptoms   Bacterial blight is typified by watery leaf spots with centers that fall out. Bacterial stem rots caused by *Erwinia* spp. are generally first noticed when young plants are transplanted. The leaves of infected plants usually yellow quickly.

Control   Control of bacterial leaf spots or blights can be best accomplished through use of clean propagation material and a watering system that either does not wet the foliage or allows it to dry rapidly. Both antibiotic and copper compounds provide little control of bacterial diseases and are not recommended. Bacterial stem rot is usually not possible to control once started. Use of clean cuttings is the only successful method of cultural control, although some growers have reported dipping in streptomycin sulfate as a moderately successful control method.

Fertilizer burn

Figure 374

Cause   High soil soluble salts caused by excessive fertilization or use of high-salt fertilizers or saline irrigation water

Signs and symptoms   Necrotic leaflet tips may appear on older leaves, yellow new foliage may emerge, and stunting is common. Roots will often have necrotic tips or more extensive necrosis. If fertilizer accumulates next to a stem, the plant may collapse at that point.

Control   Low-salt fertilizers should be used and only at recommended rates. Some leaching of the soil should occur at each irrigation. If the problem is caused by saline irrigation water and a cleaner water source cannot be found, the soil should not be allowed to dry out. If the condition already exists, the soil should be leached thoroughly several times to remove excess salts. Never pile fertilizer near the stem of any plant.

Selected Reference

Poole, R. T., and C. A. Conover. 1990. Severity of necrosis of *Spathiphyllum* 'Petite' foliage grown under various air temperatures, fertilization rates, and irrigation frequencies. University of Florida, Central Florida Research and Education Center-Apopka, CFREC-A Research Report, RH-90-18.

Magnesium deficiency

Figure 375

Cause   Insufficient magnesium (Mg) in the soil

Signs and symptoms   The oldest leaves usually have broad, chlorotic areas starting at the edges. In severe cases, the areas between the veins of such leaves may become tattered and drop out.

Control   Magnesium is readily leached from sandy soils and other soils with little cation exchange capacity. High levels of potassium or calcium in the soil also can induce Mg deficiency. Magnesium deficiency is difficult to correct once symptoms are present. It is best prevented by amending all container media with dolomite. Foliar magnesium sprays are generally ineffective in treating magnesium deficiency, since they supply very small amounts of magnesium relative to the amount required.

Myrothecium leaf spot and petiole rot

Figures 376 and 377

Cause   *Myrothecium roridum*

Signs and symptoms   Brown to black, circular lesions form on leaf margins and centers. The lesions may have concentric rings of light and dark tissue, or they may be water soaked and uniformly black. Fungal fruiting bodies, which are irregularly shaped and black and have a white fringe around the borders, are frequently found on the lower surface of the spot.

Control   Small, tissue-cultured plantlets are highly susceptible to this disease and may be lost if precautions are not taken. Iprodione provides the best chemical control of this disease but is no longer labeled for this crop because of its phytotoxicity on a single *Spathiphyllum* cultivar. Thiophanate methyl and mancozeb provide some control of Myrothecium leaf spot on *Spathiphyllum*. Disease severity is greatest during periods of the year when air temperatures are between 60 and 85°F (15 to 27°C). Little, if any, disease occurs at other times.

Selected Reference

Chase, A. R. 1983. Influence of host plant and isolate source on Myrothecium leaf spot of foliage plants. Plant Dis. 67:668-671.

Phytophthora aerial blight

Figures 378 and 379

Cause   *Phytophthora parasitica*

Signs and symptoms   Large (up to 2.5 cm wide) black or brown dead spots form on leaf margins and centers. Spots are wet and mushy under moist conditions but may dry if plant foliage is kept dry.

Control   Disease is most severe when temperatures are between 75 and 90°F (24 and 32°C). Keep plant foliage dry, and grow plants in a sterilized potting medium on raised benches away from the native soil. Soil drenches and foliar sprays with etridiazole, fosetyl aluminum, and metalaxyl are generally effective in controlling this problem. Always treat both soil and foliage, since the pathogen moves from the potting medium onto the foliage.

Selected Reference

Uchida, J. Y. 1989. Foliar blight of *Spathiphyllum* caused by *Phytophthora*. University of Hawaii, HITAHR Brief No. 084.

Southern blight

Figure 380

Cause *Sclerotium rolfsii*

Signs and symptoms Plants with southern blight may initially be similar in appearance to those infected with many other stem- or root-infecting fungi. As the disease progresses, however, the white, cottony masses of mycelia and brown, seedlike sclerotia set this disease apart. The sclerotia usually form on the basal portions of stems of infected plants but may also be found on infected leaves. Eventually the entire cutting or plant may be covered with the fungus.

Control Southern blight must be controlled through prevention, since the only effective chemical, PCNB, should be used only once per crop. Use pathogen-free potting medium, pots, and planting materials.

Syngonium

Nephthytis are popular foliage plants native to South America. The plants vine readily and are used in hanging baskets and totems, and small plants are used in dish gardens. In the past, nephthytis were grown primarily from stem cuttings called "eyes" produced in groundbeds. Many cultivars are produced from tissue-cultured plantlets; their color can range from yellow green to dark green solid and variegated forms as well as pink to burgundy shades. Plants are produced under 1,500 to 3,000 ft-c with a minimum temperature of 65°F (18°C) to maintain good growth. Nephthytis will do well with interiorscape light intensities of 75 and 150 ft-c. Nephthytis are hosts of the majority of the pathogens that affect dieffenbachias and philodendrons, including viral, bacterial, and fungal diseases of roots, stems, and leaves. Pests of this plant commonly include mealybugs and armyworms.

Acremonium leaf spot

Figure 381

Cause *Acremonium crotocinigenum*

Signs and symptoms Acremonium leaf spot of syngonium is characterized by tiny, water-soaked lesions. These spots slowly enlarge and turn reddish brown with yellow halos, which are especially prominent on the non-variegated cultivars. Spots are generally circular to slightly irregular in shape and form on the youngest leaves. Older leaves are rarely infected, although severely affected leaves frequently turn completely yellow.

Control Fungicide sprays of are of questionable value on syngonium. Control should be based on using pathogen-free cuttings, growing plants on raised benches, and minimizing exposure to rainfall and overhead irrigation.

Selected References

Alfieri, S. A., Jr., and C. Wehlburg. 1969. Cephalosporium leaf spot of *Syngonium podophyllum* Schott. Proc. Fla. State Hortic. Soc. 82:366-368.

Linn, M. B. 1940. Cephalosporium leaf spot of two aroids. Phytopathology 30:968-972.

Uchida, J. Y., and M. Aragaki. 1982. Acremonium leaf spot of *Syngonium*: Nomenclature of the causal organism and chemical control. Plant Dis. 66:421-423.

Ceratocystis or black cane rot

Figure 382

Cause *Ceratocystis fimbriata*

Signs and symptoms Brown or black, water-soaked cankers girdle stems, especially portions in contact with the potting medium. Stem lesions are sunken and elongated and usually develop at nodes. Infected plants are frequently chlorotic and stunted, and some may die. Aerial and subsurface roots show rot. Leaf lesions are dark brown to black with a bright yellow halo. The fruiting bodies (perithecia) of the pathogen frequently form on infected stems, leaves, and roots.

Control Hot water treatment of canes for 30 min at 120°F (49°C) was successful experimentally. The widespread use of tissue-cultured plantlets renders this method generally inappropriate. Other types of controls have not been investigated. However, use of pathogen-free plants and new or sterilized pots and potting media is recommended.

Selected References

Davis, L. H. 1953. Black cane rot of *Syngonium auritum*. (Abstr.) Phytopathology 43:586

Uchida, J. Y., and M. Aragaki. 1979. Ceratocystis blight of *Syngonium podophyllum*. Plant Dis. Rep. 63:1053-1056.

Cold water damage

Figure 383

Cause Cold water damage

Signs and symptoms Portions or entire leaves appear wet or water soaked. This symptom almost always occurs on young leaves, primarily in early morning during the winter when sunlight first strikes the foliage and warms the leaf tissue while the potting medium is still cold. This condition causes a temporary water imbalance in the plant; as the soil warms, symptoms disappear.

Control   To control the problem, maintain root temperature at 65°F (18°C) or above or increase air temperature slowly.

Copper toxicity

Figure 384

Cause   Foliar applications of copper-containing bactericides

Signs and symptoms   Small, water-soaked areas form at leaf tips or in low portions where water or spray accumulates. These spots resemble those caused by bacterial infections when new but generally turn tan. They are very irregular in shape.

Control   When using copper-containing bactericides or nutritional sprays, be sure to test new cultivars for sensitivity. Always use labeled rates and intervals of these products, and never mix them with a product (such as fosetyl aluminum or vinegar) that can lower the pH. Water with a pH below 6 can result in copper toxicity.

Selected Reference

Chase, A. R. 1989. Aliette 80WP and bacterial disease control. III. Phytotoxicity. University of Florida, Central Florida Research and Education Center-Apopka, CFREC-A Research Report, RH-89-9.

Erwinia blight

Figure 385

Cause   *Erwinia carotovora* subsp. *carotovora* and *E. chrysanthemi*

Signs and symptoms   Symptoms first appear as small, water-soaked lesions that can be translucent. Sometimes they are confined between leaf veins, and at other times they expand irregularly across the veins. If *Erwinia* causes leaf spot, the centers may become mushy and drop out. The spots are usually tan to dark brown, depending upon moisture conditions and activity. Tan lesions are common when the weather is dry and indicate a relatively inactive infection.

Control   Bacterial leaf spot control should be based upon use of pathogen-free cuttings or tissue-culture plantlets, since bacteria can be carried on the surface or within stems of asymptomatic plants. Minimizing overhead water is also very important because bacteria need water to spread and infect plants. Use of streptomycin sulfate or copper compounds has been moderately effective in research trials for leaf spot control when overhead irrigation and exposure to rainfall were eliminated. Some forms of copper are phytotoxic to *Syngonium,* and the form chosen should be tested by each grower under his or her specific growing conditions.

Selected References

Chase, A. R. 1987. Effect of fertilizer rate on susceptibility of *Syngonium podophyllum* 'White Butterfly' to *Erwinia chrysanthemi* or *Xanthomonas campestris.* University of Florida, Central Florida Research and Education Center-Apopka, CFREC-A Research Report, RH-87-5.

Knauss, J. F., and C. Wehlburg. 1969. The distribution and pathogenicity of *Erwinia chrysanthemi* Burkholder et al. to *Syngonium podophyllum* Schott. Proc. Fla. State Hortic. Soc. 82:370-373.

Lai, M., S. Shaffer, and K. Sims. 1978. Bacterial blight of *Syngonium podophyllum* caused by *Erwinia chrysanthemi* in California. Plant Dis. Rep. 62:298-302.

Myrothecium leaf spot and petiole rot

Figures 386 and 387

Cause   *Myrothecium roridum*

Signs and symptoms   Myrothecium leaf spot and petiole rot occur commonly on small, tissue-cultured plantlets. As few as three leaf spots can result in plant loss. Petiole rot is most common on newly established plantlets. Lesions start as small, water-soaked areas that are greasy appearing. These spots are generally circular and, when mature, contain black and white fruiting bodies on the undersides of the leaves or at petiole bases.

Control   Minimizing foliage wetting and wounding greatly reduces the severity of this disease, especially on tissue-cultured plantlets. Also, do not apply fertilizer at rates higher than necessary, since doing so can increase the plant's susceptibility to *M. roridum.* Table 15 lists the relative susceptibility of some syngoniums to Myrothecium

TABLE 15. Relative resistance of *Syngonium* cultivars to *Xanthomonas campestris* pv. *dieffenbachiae* and *Myrothecium roridum*

Degree of resistance	Cultivar resistance to *Xanthomonas*	Cultivar resistance to *Myrothecium*
Very low	Bold Allusion	Pixie II, Robust Allusion
Low	Bold, Cream Allusion, Cream Tetra, Gold Allusion, Pride	Gold Allusion, Lemon Lime, Silver Robusta
Some	Bob Allusion, Flutterby, Gold, Lemon Lime, Matisse, Pink Allusion, Pixie I, Robust Allusion, Silver Robusta, Wenlandii	Bold Allusion, Cream Allusion, Cream Tetra, Degas, Flutterby, Freckles, Maya Red, Renior, Robusta II, Roxanne, Wenlandii, White Butterfly I
Good	DaVinci, Goya, Monet, Renoir, Robusta I, White Butterfly I and II	Gold Allusion, Monet, Patricia, Pink Allusion, Pixie I, Pride, Wenlandii, Willsonii
High	Freckles, Holly, Patricia, Robusta II, White Butterfly III	Bob Allusion, Bold, Dali, Holly, White Butterfly II and III
Very high	Dali, Degas, Maya Red, Picasso, Willsonii	. . .
Nearly immune	Emerald Green (seeds), Roxanne	Emerald Green (seeds), DaVinci, Gold, Goya, Matisse, Picasso

leaf spot. Both chlorothalonil and mancozeb are very effective in controlling Myrothecium leaf spot of other foliage plants but do not provide good control of the petiole rot disease. Captan, iprodione, or triflumizole provide good control of petiole rot on syngoniums.

Selected References

Chase, A. R. 1990. Controlling Myrothecium petiole rot of *Syngonium podophyllum*. Proc. Fla. State Hortic. Soc. 103:194-195.

Chase, A. R. 1993. Are your syngoniums resistant to disease? University of Florida, Central, Florida Research and Education Center-Apopka, CFREC-A Research Report, RH-93-19.

Rhizoctonia aerial blight and leaf spot

Figure 388

Cause   *Rhizoctonia solani*

Signs and symptoms   Rhizoctonia aerial blight of *Syngonium* usually appears as small, irregularly shaped, water-soaked lesions on lower leaves or leaf edges in contact with the potting medium. Lesions are brown and may be accompanied by the weblike mycelium of the pathogen, which is also reddish brown.

Control   The fungicides mentioned for control of Myrothecium leaf spot are also effective for control of Rhizoctonia aerial blight. Since the pathogen is soilborne, plant roots must be treated for optimal disease control. For this reason, thiophanate methyl or iprodione are frequently used, since the other compounds are not usually recommended for drench applications. Always use pathogen-free potting media, pots, and plants, and grow plants on raised benches to avoid infections. Since the disease occurs during the hot, humid summer months, special precautions should be taken at that time to prevent infections.

Selected Reference

Knauss, J. F. 1973. Rhizoctonia blight of *Syngonium*. Proc. Fla. State Hortic. Soc. 86:421-424.

Xanthomonas blight

Figures 389, 390, and 391

Cause   *Xanthomonas campestris* pv. *dieffenbachiae*

Signs and symptoms   Xanthomonas blight symptoms on *Syngonium* occur first on the leaf margins where the bacterium enters through hydathodes. Lesions are at first translucent, yellowish, and water soaked. They may take a long time to enlarge, but eventually they can encompass the entire leaf margin, invade the center of the leaf, and even cause leaf abscission. Mature lesions are papery and tan and can be surrounded by a bright yellow halo. If the plant becomes systemically infected, it will show signs of yellowing, stunting, and loss of lower leaves. Eventually systemically infected plants die.

Control   Use of bactericides for control of even the foliar phase of Xanthomonas blight is rarely effective. Avoidance of this disease is the most effective control. Scout the crop routinely and frequently to detect early symptoms of Xanthomonas blight. Table 15 lists the relative susceptibility of some syngoniums to Xanthomonas blight. Use of white vinegar (1% by volume) has occasionally been effective in reducing disease spread. Copper compounds are as effective, although no bactericides can control this disease once plants become systemically infected. In addition, fosetyl aluminum has been shown to reduce disease development and spread when applications are made prior to infection. Limit overhead irrigation to reduce pathogen spread, and keep in mind that most of the commonly produced aroids (dieffenbachia, aglaonema, and anthurium) are also hosts of this pathogen.

Selected References

Chase, A. R. 1988. Effects of temperature and preinoculation light level on severity of Syngonium blight caused by *Xanthomonas campestris*. J. Environ. Hort. 6:61-63.

Chase, A. R. 1989. Effect of nitrogen and potassium fertilizer rates on severity of Xanthomonas blight of *Syngonium podophyllum*. Plant Dis. 73:972-975.

Chase, A. R. 1993. Are your syngoniums resistant to disease? University of Florida, Central Florida Research and Education Center-Apopka, CFREC-A Research Report, RH-93-19.

Chase, A. R., P. S. Randhawa, and R. H. Lawson. 1988. New disease of *Syngonium podophyllum* 'White Butterfly' caused by a pathovar of *Xanthomonas campestris*. Plant Dis. 72:74-78.

Dickey, R. S., and C. H. Zumoff. 1987. Bacterial leaf blight of *Syngonium* caused by a pathovar of *Xanthomonas campestris*. Phytopathology 77:1257-1262.

Tolmeia (Piggyback plant)

The piggyback plant is native to western North America. Plants are produced with 3,500 to 4,000 ft-c. Few diseases have been found on piggyback plant, although mealybugs, whiteflies, and spider mites can be quite troublesome.

Anthracnose

Figure 392

Cause   *Colletotrichum gloeosporioides*

Signs and symptoms   Leaf spots are initially tan and 1 to 2 mm in diameter. Under optimal conditions, the spots merge and form large, dead spots across the leaf. The fruiting bodies of the pathogen may appear in concentric rings of tiny black specks on the upper surface.

Control   Keep foliage as dry as possible by watering early in the day, using fans, and reducing relative humidity by ventilation at sundown. Disease development is greatest at temperatures between 60 and 80°F (15 and 27°C). Mancozeb and chlorothalonil provided excellent disease control experimentally.

Selected Reference

Pierce, L., and A. H. McCain. 1990. Anthracnose of piggyback plant caused by *Colletotrichum gloeosporioides* (Penz.) Sacc. J. Environ. Hort. 8:207-209.

Tradescantia (Wandering jew)

Wandering jews are native to North and South America and are used primarily in hanging baskets and as ground covers. They can be produced under 3,500 to 4,500 ft-c and a minimum temperature of 50°F (13°C). Wandering jews have few important diseases but are affected by mealybugs and scale insects.

Pythium root rot

Figure 393

Cause   *Pythium splendens* and other *Pythium* species

Signs and symptoms   Root rot is typified by wilting of plants and yellowing of the lower leaves. Infected roots are brown to black, reduced in mass, and mushy. The outer portion of infected roots can easily be pulled away from the inner core.

Control   Using pathogen-free potting medium and pots and growing plants on raised benches can eliminate much of this problem. If fungicides are needed, drenches with etridiazole, fosetyl aluminum, or metalaxyl can aid in control of Pythium root rot. Since many times other pathogens are also involved, accurate diagnosis of the cause must be made prior to choice of fungicides. As usual, be sure to check labels for legal uses on *Tradescantia*.

Tradescantia mosaic

Figure 394

Cause   Tradescantia mosaic virus

Signs and symptoms   Leaf deformity and stunting occur.

Control   Discard all plants with these symptoms, since any cuttings removed from them are likely to have the virus. *Zebrina* species are also hosts of this virus. It is known to be transmitted both mechanically and by aphid vectors.

Selected References

Baker, C. A., and F. W. Zettler. 1988. Viruses infecting wild and cultivated species of the Commelinaceae. Plant Dis. 72:513-518.
Lockhart, B. E., J. A. Betzold, and F. L. Pfleger. 1981. Characterization of a potyvirus causing a leaf distortion disease of *Tradescantia* and *Zebrina* species. Phytopathology 71:602-604.

Xanthosoma

A few species of the genus *Xanthosoma* are grown as specimen plants in the foliage industry. They should be produced at temperatures higher than 59°F (15°C) to avoid chilling injury. *Xanthosoma* spp. are subject to a serious bacterial disease as well as to a variety of fungal diseases.

Myrothecium leaf spot and petiole rot

Figures 395 and 396

Cause   *Myrothecium roridum*

Signs and symptoms   Myrothecium leaf spot most frequently appears on wounded areas of leaves, such as tips and breaks in the main vein that occur during handling. The leaf spots are watery and nearly always contain the black and white fungal fruiting bodies in concentric rings near the outer edge of the spot on the leaf undersides. The presence of these bodies is good evidence that the cause is *Myrothecium*. Newly planted, tissue-cultured explants are especially susceptible to this disease. The primary symptom on these explants is petiole rot starting with the oldest leaves, although leaf spot can occur as well.

Control   Many other plants are hosts of *M. roridum*, including *Aglaonema, Aphelandra, Begonia, Calathea, Spathiphyllum,* and *Syngonium* spp., and these plants must be included in control programs. Avoid wounding leaves, and keep the foliage as dry as possible. Iprodione is very effective in controlling *Myrothecium* on tissue-cultured plantlets affected with petiole rot. Preventive treatments to newly transplanted tissue-cultured plants are recommended. In addition, mancozeb provides good control of the leaf spot on some hosts. Check labels for legal uses on *Xanthosoma*.

Xanthomonas leaf spot

Figure 397

Cause   *Xanthomonas campestris* pv. *dieffenbachiae*

Signs and symptoms   Foliar infections on xanthosomas start as pinpoint, water-soaked areas that can rapidly enlarge to 7 mm or more. They tend to form on leaf margins where the bacterium can enter the leaf through hydathodes. When they invade a main vein in the leaf, the infection rapidly spreads throughout the leaf. These necrotic areas are frequently very black and surrounded by a bright yellow halo. Most other plants in the Aroid family, such as *Aglaonema, Anthurium,* and *Syngonium* spp., are also hosts of this pathogen.

Control   Eliminate all stock plants that have Xanthomonas leaf spot. The disease is very difficult to control unless plants are produced without overhead watering or exposure to rainfall. Bactericides such as copper-containing compounds may be somewhat effective if used on a preventative and regular basis. Nutritional studies on dieffenbachias have shown that applications of fertilizer at rates greater than recommended result in decreased susceptibility to *X. campestris* pv. *dieffenbachiae*.

Yucca

Selected References

Berniac, M. 1974. Une maladie bacterienne de *Xanthosoma sagittifolium* (L.) Schott. Ann. Phytopathol. 6:197-202.

Pohronezny, K., W. Dankers, M. Lamberts, and R. B. Volin. 1986. Bacterial leaf spot of malanga (*Xanthosoma caracu*): A bacterial disease of potential importance to the tropical foliage industry. Proc. Fla. State Hortic. Soc. 99:325-328.

Yucca

Yucca species are native to semiarid or well-drained areas of North America. Many of the species reach 10 m in height and have a woody stem. Indoor use is usually restricted to those varieties without a spine at the leaf tip. Yuccas are propagated from cane sections produced under full sun; finished plants are grown under 6,000 to 7,000 ft-c or full sun. Interior light levels of at least 150 ft-c are needed to grow this plant. Temperatures between 65 and 95°F (18 and 35°C) allow good growth of yuccas. Yuccas are commonly infected with fungi that cause leaf spot diseases. Mealybugs, scales, and cane weevils also may be pests of yuccas.

Cercospora leaf spot

Figure 398

Cause   *Cercospora* sp.

Signs and symptoms   Initially rust-colored specks form. The spots enlarge into elliptical areas that are up to 7 mm across and become tan to brown.

Control   Protect plants from overhead irrigation and rainfall if possible to diminish the spread of spores and the conditions needed for them to germinate and infect leaves. Remove leaves with symptoms, and avoid using cuttings that have spots when they are stuck. No fungicides have been evaluated for control of this disease on yucca.

Coniothyrium or brown leaf spot

Figure 399

Cause   *Microsphaeropsis concentrica* (= *Coniothyrium concentricum*)

Signs and symptoms   Coniothyrium leaf spot first appears as tiny, clear zones in older leaves of yucca. Lesions turn yellowish and finally brown as they mature. They are generally elliptical and are scattered across the entire upper surface of older leaves. After about 4 months, black perithecia or pycnidia form in sunken lesion centers and are easily seen with the naked eye. Sometimes a yellow border forms around each spot. A dark purple or black margin forms around older lesions, which rarely exceed 7 mm in diameter. Although it appears that lesions occur on older leaves only, this is probably because of the length of time needed for a lesion to develop rather than any difference in susceptibility. This disease has been reported on many species of *Yucca* including *Y. aloifolia*, *Y. filamentosa* (Adam's-needle), and *Y. smalliana*.

Control   One of the most important aspects of control is removal of older infected leaves and elimination of overhead watering or rainfall. Since most yuccas are grown exposed to both overhead watering and rainfall, regular applications of fungicides are needed to prevent disease. Weekly applications of chlorothalonil, zineb, or mancozeb provided excellent disease control experimentally.

Selected References

Chase, A. R. 1984. Controlling brown leaf spot of *Yucca*. Florida Nurseryman 31(3):61.

Sobers, E. K. 1967. The perfect stage of *Coniothyrium concentricum* on leaves of *Yucca aloifolia*. Phytopathology 57:234-235.

Cytosporina or gray leaf spot

Figure 400

Cause   *Cytosporina* sp.

Signs and symptoms   Cytosporina or gray leaf spot of *Yucca aloifolia* (Spanish-bayonet) has not been described in depth. Symptoms include tip and marginal necrosis with a predominantly gray color and a brown margin. Lesions often reach 7.5 cm, have concentric rings of light and dark tissues, and are most common on older leaves. The pycnidia of *Cytosporina* sp. can be found in these lesions.

Control   Use the methods and fungicides described for Coniothyrium leaf spot to control Cytosporina leaf spot of yuccas.

Selected Reference

Wehlburg, C. 1969. Two leaf spot diseases of Spanish bayonet. Fla. Dept. of Agric., Div. of Pl. Indus., Pl. Pathol. Circ. No. 79.

Fusarium stem rot

Figures 401 and 402

Cause   *Fusarium* spp., *Nectria* spp.

Signs and symptoms   One of the most common diseases of rooted yucca cuttings is caused by *Fusarium* spp. Soft rot of stem ends develops, and leaves are destroyed as well. The pathogen commonly produces two types of fruiting bodies on infected tissue: bright red perithecia, which are round and relatively easy to see with the naked eye, and powdery clusters of conidia, which are ochre-colored.

Control   Use only pathogen-free cuttings and new or sterilized pots and potting media, and grow plants on raised benches. No fungicides have been evaluated for control of Fusarium stem rot on yucca.

Southern blight

Figure 403

Cause   *Sclerotium rolfsii*

85

Signs and symptoms   *Sclerotium rolfsii* attacks all portions of the plant but is most commonly found on stems. White, relatively coarse mycelia grow in a fanlike pattern and may be seen on the soil surface or stems. The round sclerotia form almost anywhere on the plant or soil surface. Sclerotia are initially white and cottony and approximately the size of a mustard seed. As sclerotia mature, they turn tan and eventually dark brown and harden.

Control   Although this disease can be avoided through cultural methods, it continues to cause losses in production of foliage plants today. Chemical control of southern blight has been investigated on several foliage plants as well as on nonornamental crops. PCNB provides the most efficacious control of southern blight on many foliage plants. Applications of this fungicide to some foliage plants results in stunting, and its use is restricted to once every 12 months.

Glossary

C—centigrade or Celsius

cm—centimeter (0.39 in.; 10 mm = 1 cm)

F—Fahrenheit

ft—feet

ft²—square feet

ft-c—foot candle

g—gram

gal—gallon

in.—inch (2.54 cm)

kg—kilogram (2.23 lb)

L—liter (1.06 quarts)

lb—pound (453.59 g)

m—meter (39.37 in.)

mg—milligram

ml—milliliter (1 ml=0.001 L)

mm—millimeter (0.04 in.; 10 mm=1 cm)

oz—ounce (28.34 g) or fluid ounce (29.6 ml)

ppm—parts per million

abiotic—not living

abscise (n. abscission)—to fall off, as with leaves, flowers fruits, or plant parts

acaricide—a pesticide used to control mites and ticks; miticide

acervulus (pl. acervuli)—erumpent, saucer-shaped, cushionlike fruiting body of a fungus bearing conidiophores, conidia, and sometimes setae

active ingredient (a.i.)—the substance in a pesticide that kills or controls a pest

actual dosage—the amount of active ingredient that is applied to an area

acute—pertaining to symptoms that develop suddenly, as opposed to chronic

additive—*see* adjuvant

adherence—the act or ability of a substance to stick to a surface

adjuvant—any substance added to a pesticide formulation to make the active ingredient work better (e.g., an adhesive, emulsifier, penetrant, spreader, or wetting agent)

aerobic—living, active, or occurring only in the presence of oxygen

aerosol—a suspension of colloidal particles in air or gas

anatomy—the branch of plant morphology that deals with the internal structure and form of plants

antagonism—the decrease in effectiveness resulting from two or more chemicals being exposed to each other or mixed together

anthracnose—a disease with limited necrotic lesions, caused by fungi that produce spores borne in acervuli

antibiotic—damaging to life; a chemical, usually of microbial origin, that inhibits or kills other microorganisms

antitranspirant—a chemical that reduces water loss (i.e., prevents drying out)

aphid—insect (Homoptera) that feeds on juices of many types of plants, possibly causing wilting, distorted growth, or gall formation, and that may serve as the vector of certain viral diseases of plants

asexual—vegetative; without sex organs, gametes, or sexual spores; imperfect

atomize—to reduce a liquid or solid to fine droplets or particles

avirulent—unable to cause disease; nonpathogenic

bactericide—a pesticide used to control bacteria

bacterium (pl. bacteria)—minute, prokaryotic organism that usually lacks chlorophyll and exists mostly as a parasite or saprophyte

basal—located at or near the base of a structure

biological control—disease or pest control through counterbalance of microorganisms and other natural components of the environment

blight—sudden, severe, and extensive wilting and/or death of leaves, stems, flowers, or entire plants

blotch—irregularly shaped, usually superficial spot or blot

broad-spectrum pesticide—a pesticide that controls or is toxic to a wide range of pests; nonselective pesticide

canker—a stem lesion with sharply limited necrosis of the cortical tissue

canopy—expanded leafy top of a plant or plants

carbamate—a synthetic organic pesticide (acaricide, fungicide, herbicide, insecticide, or nematicide) that contains carbon, hydrogen, nitrogen, and sulfur and belongs to a group of chemicals that are salts or esters of carbonic acid

causal agent—anything (biotic or abiotic) capable of causing a disease

chemical control—control method based on use of chemicals to reduce or eliminate a pest population

chemical name—the scientific name that describes the contents or formula of the active ingredients of a pesticide

chlorophyll—green, light-sensitive pigments, found chiefly in chloroplasts of higher plants, that participate in photosynthesis

chlorosis (adj. chlorotic)—failure of chlorophyll development caused by a nutritional disturbance or disease; fading of green plant color to light green, yellow, or white

cladophyll—stem or leaf section of certain cacti (*Schlumbergera* spp.) used in propagation

clone—one of a group of genetically identical individuals resulting from asexual (vegetative) multiplication; any plant propagated vegetatively and therefore considered a genetic duplicate of its parent

coalesce—to grow together, overlap, merge

common name (generic name)—the well-known, simple name of a pesticide accepted by the Pesticide Regulation Division of the EPA

compatibility—the ability of two or more pesticides to be chemically and physically mixed without altering their effectiveness

concentration—the amount of ingredient contained in a given volume or weight

conidium (pl. conidia)—asexual spores produced by fungi

cortex (adj. cortical)—tissues between the epidermis and phloem in stems and roots

crown—compacted series of nodes from which shoots and roots arise

crown rot—type of disease or symptom in which the crown is rotted

cultivar—cultivated variety; group of closely related plants of common origin within a species that differ from other cultivars in certain minor details such as form, color, flower, or fruit

cultural control—methods that do not involve use of chemical or biological techniques but result in pest control (e.g., sanitation, irrigation management, and weed control)

culture—artificial growth and propagation of organisms on nutrient media or living plants

culture index—culture of stem sections on defined medium to test for presence of fungal or bacterial vascular pathogens

curative—relating to or used in the cure of diseases

cuticle—outer sheath or membrane of a plant

damping-off—rapid, lethal decline of germinating seed or seedlings before or after emergence

decline—condition characterized by general reduced vigor, dwarfed leaves and shoots, chlorosis, wilting, leaf drop, and reduced flower quality

desiccate—to dry up

diagnosis—identification of the nature and cause of a plant malady (abnormality or malfunction)

dieback—progressive death of leaves, stems, or roots from the tips back

dip—complete or partial immersion of a plant in a pesticide

disease—any abnormal, malfunctioning process in the host induced by the constant association of one or more causal agents

disease complex—diseases resulting from combined or sequential actions of two or more biotic or abiotic agents

disinfectant—an agent that kills or inactivates organisms present within the plant tissue

disinfestant—an agent that kills or inactivates organisms present on the surface of the plant or plant part or in the immediate environment

dissemination—spread of infectious material (inoculum) from a diseased to a healthy plant by wind, water, people, animals, insects, mites, machinery, etc.

dormant—resting; living in a state of reduced physiologic activity

drench—an application method used to wet a material thoroughly; a liquid chemical substance poured onto the soil around a plant rather than applied as a foliar spray

edema—swelling symptom caused by excessive moisture that appears as numerous small bumps on the lower sides of leaves or on stems

epidemic—general and serious outbreak of disease (used loosely of plants)

epidermis—superficial layer of cells on all plant parts

epinasty—abnormal twisting and bending of stems and downward bending of leaves

eradication—control of disease by eliminating the pathogen after it has become established

erumpent—bursting or erupting through the surface, as a rust pustule

ethylene—colorless, odorless gas, often emitted by fruit, foliage, and incomplete combustion of oil and gas in heaters, that hastens the senescence of flowers

etiolation—yellowing of tissue and elongating of stems caused by reduced light

etiology—study of the cause or origin of a disease

exudate—that which is excreted or discharged; ooze

FIFRA (Amended)—Federal Insecticide, Fungicide, and Rodenticide Act (Amended), the federal law pertaining to pesticide regulations and use in the United States; often includes (Amended) because the original act has been changed

fleck—minute spot

fogger—an aerosol generator; a piece of pesticide equipment that breaks some pesticides into very fine droplets (aerosols or smokes) and blows or drifts the "fog" onto the target area

foliar—pertaining to leaves

foliar application—spraying a pesticide onto the stem, leaves, needles, and blades of grasses, plants, shrubs, or trees

foot rot—basal rot of a plant usually affecting stems and crowns

formulation—the physical nature (i.e., EC, S, WP, etc.) of a pesticide product (it may contain one or more active ingredients, the carrier, and other additives)

fruiting body—any complex, spore-bearing fungal structure, e.g., acervuli and pycnidia

fumigant—vapor-active chemical used against microorganisms and other pests

fumigate—to apply a vapor-active (volatile) disinfectant to kill microorganisms and other pests

fungicide (adj. fungicidal)—chemical or physical agent that kills or inhibits the growth of fungi

fungus (pl. fungi)—organism lacking chlorophyll that reproduces by sexual or asexual spores and not by fission; generally speaking, a mycelium with well-marked nuclei

gall—abnormal swelling or localized outgrowth, often more or less spherical, produced by a plant as the result of attack by a fungus, bacterium, insect, mite, or other agent

genus—taxonomic category ranking above species and below family (the generic name of an organism is the first word of the binomial)

germinate—to begin growth, as of a seed, spore, sclerotium, or other reproductive body

girdle—to encircle

globose—nearly spherical

herbaceous—pertaining to plants that do not develop much woody tissue and thus remain soft and succulent

herbicide—chemical or physical agent that limits the growth of or kills plants

hormone—organic chemical normally produced in minute amounts in one part of an organism and transported to another area of the same organism, where it affects growth and/or other functions

host—living plant attacked by (or harboring) a living parasite and from which the invader obtains part or all of its nourishment

host range—kinds of plants attacked by a given pathogen

hybrid (v. hybridize)—offspring of two individuals of different genetic character

hydathode—epidermal structure specialized for secretion or exudation of water

hypersensitive—displaying increased sensitivity, as when host tissue dies at the point of attack by a pathogen so that infection does not spread

hypha (pl. hyphae)—tubular filament of a fungal thallus or mycelium

immune—a state of not being affected by a disease or poison; totally resistant

in vitro—in glass or an artificial environment

in vivo—within a living organism

incidence—the degree or range of occurrence

incompatible—pertaining to the attempted mixture of two or more pesticides that fails because of adverse chemical or physical reaction or loss in the effectiveness of one or more of the products

inert ingredient—the substance in a pesticide product that has no pesticidal (controlling or killing) action; an inactive ingredient

infection—the establishment of a pathogen or parasite within the host, resulting in disease

infectious—capable of infecting and spreading from plant to plant

infest—to contaminate, as with organisms

infestation—attack by animals, especially insects or nematodes; aggregation of inoculum or other organisms on a plant surface

injury—result of transitory operation of an adverse factor such as insect feeding, action of a chemical, or adverse environmental factor

insecticide—chemical or physical agent used to control or kill insects

integrated control—the use of more than one approach to or method of pest control (includes cultural, biological, and chemical practices)

internode—area between two adjacent nodes on a stem

interveinal—area between leaf veins

isolate—separated or confined spore or microbial culture; a fungus, bacterium, or other organism in pure culture

label—technical information about the pesticide in the form of printed material attached to or printed on the pesticide container

latent—present but not manifested or visible

leach—to wash soluble nutrients down through the soil

leaf spot—self-limiting or localized lesion on a leaf

lesion—well-marked but localized diseased area

lethal—causing death; fatal; deadly

mechanical transmission or **inoculation**—spread or introduction of inoculum to infection courts (especially wounds) accompanied by physical disruption of host tissues

microscopic—too small to be seen without the aid of a microscope

midrib—central, thickened vein of a leaf

mildew—plant disease characterized by a thin coating of mycelial growth and spores on the surfaces of infected plant parts

mites—tiny animals that are related to insects and have eight jointed legs, two body regions, and no antennae (feelers) or wings

miticide—chemical or physical agent that kills or inhibits the growth of mites; acaricide

mold—any profuse fungal growth

mosaic—disease symptom caused by disarrangement or unequal development of the chlorophyll content and characterized by a mottling of the foliage or by variegated patterns of dark and light green to yellow that form a mosaic

multiple infection—invasion by more than one parasite

mycelium (pl. mycelia)—mass of hyphae that comprises the thallus or body of a fungus

necrosis (adj. necrotic)—death, usually accompanied by darkening or discoloration

nematicide—chemical or physical agent that kills or inhibits nematodes

nematode—small, wormlike animal, parasitic in plants or animals or free-living in soil or water

node—region on the stem where a leaf is attached; the point of branching of the stem

nontarget—pertaining to any plant, animal, or other organism that is not the object of a pesticide application

obligate parasite—organism that can survive only on or in living tissue and that has not been cultured on laboratory media

ornamental (ornamental plant)—plant grown for beautification purposes such as flowers, shrubs, and trees that adorn homes, parks, and city streets (may be woody or herbaceous)

overwinter—to survive the winter (sometimes used to describe carrying over from one crop to the next)

parasite (adj. parasitic)—organism living in or on another living organism (host) and obtaining food from it

pasteurization—method of destroying a selective microbial population by heating to a prescribed temperature for a specified period of time

pathogen (adj. pathogenic)—organism or agent that causes disease in another organism

pathogenicity—ability to cause disease

pathovar (pv.)—a strain or set of strains with the same or similar characteristics, differentiated at subspecific level from other strains of the same species or subspecies on the basis of distinctive pathogenicity to one or more plant hosts

peat—partially decomposed plant tissue formed in water of marshes, bogs, or swamps, usually under conditions of high acidity

perfect stage (state)—stage in the life cycle of fungi in which sexual spores (e.g., ascospores and basidiospores) are formed after nuclear fission; sexual stage

perithecium (pl. perithecia)—flask-shaped or subglobose fungal fruiting body containing ascospores

pest—any organism injurious to plants or plant products

pesticide—a chemical substance, compound, or other agent (bactericide, fungicide, herbicide, insecticide, miticide, or nematicide) used to control, destroy, prevent damage by, or protect something from a pest

petiole—the stem of a leaf; the stalk attaching a leaf blade to a stem

pH—measure of acidity (pH 7 is neutral; below pH 7 is acidic; above pH 7 is alkaline)

photosynthesis—manufacture of carbohydrates from carbon dioxide and water in the presence of chlorophyll, using light energy and releasing oxygen

phytopathology—plant pathology; the science of plant disease

phytotoxic—harmful to plants

predispose—to make prone to infection and disease

propagation—reproduction by sexual or vegetative (asexual) means

propagule—any part of an organism capable of independent growth

protectant—agent, usually a chemical, that prevents or inhibits infection

prune—to remove stems or branches of woody plants to control size and shape and improve quality and/or quantity of fruit and flowers

pseudomonad—bacterium of the genus *Pseudomonas*

pycnidium (pl. pycnidia)—asexual, flask-shaped or globose fungal fruiting body containing conidia (pycnidiospores)

quarantine—legislative control of the transport of plants or plant parts to prevent disease spread; law

recommendation—a suggestion from or advice given by a farm advisor, extension specialist, county agent, or other agricultural authority or specialist

registered pesticide—a pesticide approved by the U. S. Environmental Protection Agency for use as stated on the label of the container

resistance—ability of a host plant to overcome completely or to suppress, prevent, or impede the activity of a pathogen

restricted use pesticide—a pesticide that has been classified, under provisions of FIFRA (Amended), for use only by an appropriately certified applicator

rhizome—jointed, underground stem

ring spot—disease symptom characterized by yellowish or necrotic rings with green tissue inside the ring, as in some plant diseases caused by viruses

rogue—to remove and destroy by hand individual plants that are diseased, infested by insects, or otherwise undesirable

rot—softening and disintegration of succulent plant tissue as a result of fungal or bacterial infection

salinity—the relative concentration of salts (especially sodium chloride) in water or soil

sanitation—destruction of infected and infested plants or plant parts

saprophyte (adj. saprophytic)—organism that uses nonliving organic matter as food

sclerotium (pl. sclerotia)—hard, usually darkened and rounded mass of dormant hyphae

sign—indication of disease from direct visibility of the pathogen or its part

site—an area, location, building, structure, plant, animal, or other organism to be treated with a pesticide to protect it from or to reach and control the target pest

slurry—a watery mixture, such as liquid mud or cement; (fungicides can be applied to seeds as slurries to minimize dustiness and improve adherence)

solution—the mixture of one or more substances into another substance (usually a liquid) in which all ingredients are completely dissolved without their chemical characteristics changing (this mixture will not settle out or separate during normal use)

sooty mold—dark fungus usually growing in insect honeydew or other high-carbohydrate substance

sp. (pl. **spp.**)—species (a genus name followed by sp. means that the particular species is undetermined; spp. after a genus name means that several species are being referred to without being named individually)

spathe—a large, leaflike part or pair of such parts enclosing a flower cluster

spore—one- to many-celled reproductive body of fungi and lower plants

sporodochium (pl. sporodochia)—superficial, cushion-shaped, asexual fruiting body

sporulate—to produce spores

spot—limited, chlorotic or necrotic, circular to oval area on leaves or other plant parts

spreader—an adjuvant that reduces the surface tension and generally increases the area that a given volume of liquid will cover on a solid surface (such as a leaf)

sterile—infertile; free from contaminant organisms

sterilization—method of destroying all microorganisms by heating to 100°C for 20 minutes, generally using free steam

sticker—an adjuvant (extender) that increases the adherence of a pesticide

stock—portion of the stem and associated root system onto which a scion is grafted; artificial breeding group; a production planting

strain—biotype; race; an organism or group of organisms that differs in minor aspects from other organisms of the same species or variety

streak—necrosis along vascular bundles in leaves or stems of grasses

stunted—unthrifty; reduced in size and vigor because of unfavorable environmental conditions or a wide range of pathogens or abiotic agents

surfactant—monomolecular compound used as a detergent that reduces surface tensions and provides spreading action when used with pesticides

susceptible—not immune; lacking resistance; prone to infection

symptom—indication of disease by reaction of the host

symptomatology—study of disease symptoms

systemic—pertaining to chemicals or pathogens (or single infections) that spread generally throughout the plant body as opposed to remaining localized

target—an area, building, animal, plant, or pest that is to be treated with a pesticide

tissue analysis—analysis of leaf tissues for major and minor elements

tissue culture—the technique of cultivating cells, tissues, or organs in a sterile, synthetic medium

tolerant—capable of sustaining disease without serious damage or yield loss

toxicity—capacity of a substance to produce injury

translocation—distribution of a pesticide from the point of absorption to other parts of the plant or animal

translucent—so clear that light rays may pass through

transmission—spread of virus or other pathogen from plant to plant

transparent—so clear that bodies may be seen through it

transpiration—loss of water vapor from aerial parts of plants, chiefly through stomata in the leaves

variety—group of closely related plants of common origin within a species that differ from each other in certain minor details such as form, color, flower, and fruit

vascular—pertaining to conductive (xylem and phloem) tissue

vector—agent (insect, mite, animal, human, etc.) able to transmit a pathogen (virus, bacterium, fungus, phytoplasma, or nematode)

viability (adj. viable)—state of being alive; ability of seeds, fungal spores, sclerotia, etc. to germinate

virulence (adj. virulent)—degree of pathogenicity; capacity to cause disease

virus—submicroscopic, filterable agent that causes disease and multiplies only in living cells and contains nucleic acid surrounded by a protein coat

virus index—assay of plant tissues for presence of virus

water-soaked—pertaining to plants or lesions that appear wet, dark, and usually sunken and translucent

whorl—circle of leaves or flowers arising from one point

wilt—lack of freshness or drooping of leaves from lack of water (inadequate water supply or excessive transpiration); a vascular disease that interrupts the normal uptake and distribution of water

wound—an injury to a plant caused by cutting, scraping, or other external force

xanthomonad—bacterium of the genus *Xanthomonas*

Color Figures

1. Botrytis blight of *Aeschynanthus*.

2. Corynespora leaf spot of *Aeschynanthus*.

3. Myrothecium leaf spot of *Aeschynanthus*.

4. Rhizoctonia cutting rot of *Aeschynanthus*.

5. Rhizoctonia aerial blight of *Aeschynanthus*.

6. Rhizoctonia dieback of *Aeschynanthus*.

7. Anthracnose on *Aglaonema*.

8. Bent tip on *Aglaonema* 'Silver Queen.'

9. Bent tip disorder on *Aglaonema*.

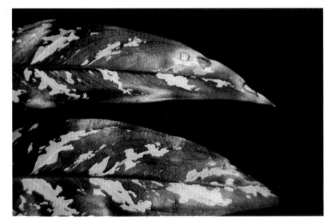

10. Boron toxicity on *Aglaonema* 'Maria.'

11. Chilling injury of *Aglaonema* 'Silver Queen.'

12. Copper deficiency on *Aglaonema*.

13. Copper toxicity on *Aglaonema*.

14. Dasheen mosaic of *Aglaonema*.

15. Erwinia leaf spot of *Aglaonema*.

16. Erwinia cutting rot of *Aglaonema*.

17. Erwinia basal rot of *Aglaonema*.

18. Fluoride toxicity of *Aglaonema* 'Silver Queen.'

19. Fusarium stem rot of *Aglaonema.*

20. Myrothecium leaf spot of *Aglaonema.*

21. *Aglaonema* with Pythium root rot (right) and healthy cutting (left).

22. Yellowing and stunting of *Aglaonema* caused by lack of fertilizer.

23. Xanthomonas blight on *Aglaonema.*

24. Anthracnose on spadix of *Anthurium*.

25. Chimera on *Anthurium* leaf.

26. Copper toxicity from a bactericide application to *Anthurium*.

27. Impatiens necrotic spot on *Anthurium*.

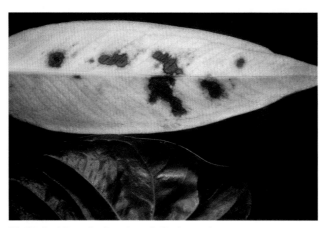

28. Phytophthora leaf spot on *Anthurium scherzeranum*.

29. Phytophthora flower blight on *Anthurium scherzeranum*.

30. Pseudomonas leaf spot on *Anthurium*.

31. Pythium root rot on *Anthurium scherzeranum*.

32. Xanthomonas blight of *Anthurium andraeanum*.

33. Systemically infected anthuriums stunted by Xanthomonas blight (left) and a single healthy plant (right).

34. Bendiocarb toxicity on *Aphelandra*. (Courtesy J. M. F. Yuen)

35. Chlorosis on *Aphelandra* caused by nitrogen deficiency. (Courtesy J. M. F. Yuen)

36. Corynespora leaf spot on *Aphelandra*.

37. Crinkle leaf disorder on *Aphelandra squarrosa* 'Apollo' (bottom) and normal plants (top).

38. Kutilakesa stem canker of *Aphelandra*. (Courtesy Florida Foliage Association)

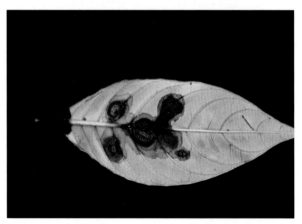

39. Myrothecium leaf spot on *Aphelandra squarrosa* 'Dania.'

100

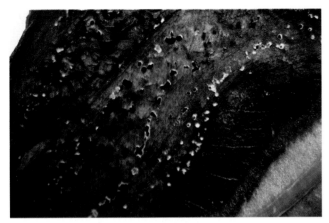

40. Myrothecium leaf spot on *Aphelandra* with sporulation.

41. Phytophthora stem rot on *Aphelandra* (plant on the lower left is healthy).

42. Tomato spotted wilt on *Aphelandra*. (Courtesy L. Barnes)

43. Anthracnose (needle necrosis) on *Araucaria*.

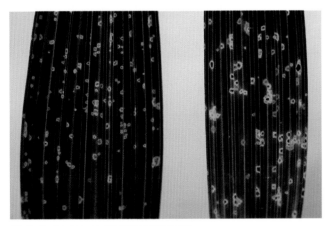

44. Anthracnose on *Washingtonia* palm.

45. Boron deficiency on *Chrysalidocarpus*.

46. Calonectria leaf spot on *Kentia* palm.

48. Damping-off of *Chrysalidocarpus* seedlings.

47. Copper toxicity on *Chamaedorea*.

49. Fertilizer toxicity on palm.

50. Fluoride toxicity on *Chrysalidocarpus*.

51. Helminthosporium leaf spot on *Chrysali-docarpus* caused by *Bipolaris setariae*.

52. Helminthosporium leaf spot on *Chrysali-docarpus* caused by *Exserohilum rostratum*.

53. Iron toxicity on *Chrysalidocarpus*.

54. Magnesium deficiency on *Phoenix*.

55. Phytophthora aerial blight on *Chamaedorea*.

56. Phytophthora root rot on *Kentia*.

57. Pink rot on *Chamaedorea*.

59. Pseudocercospora leaf spot on *Rhapis*.

58. Potassium deficiency on *Phoenix*.

60. Pseudomonas blight on *Caryota.*

61. Sclerotinia blight on *Chrysalidocarpus.*

62. Sunburn on *Chamaedorea.*

63. Zinc deficiency on *Chrysalidocarpus.*

64. Botrytis blight on begonia cuttings.

65. Fusarium stem rot on begonia.

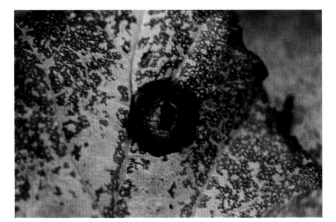

66. Myrothecium leaf spot on Rex begonia.

67. Myrothecium stem rot and dieback on Rex begonia.

68. Powdery mildew on wax begonia.

69. Pythium root rot on Rex begonia (plant on the left is healthy). (Courtesy L. S. Osborne)

70. Xanthomonas blight on Rex begonia.

71. Desiccation damage on *Vriesea*.

72. Erwinia basal stem rot on *Vriesea*.

73. Erwinia whorl rot on bromeliad. (Courtesy G. W. Simone)

74. Fusarium rot on *Tillandsia*.

75. Helminthosporium leaf spot on *Aechmea*. (Courtesy J. M. F. Yuen)

76. Pythium root rot on *Tillandsia*.

77. Pythium root rot on *Aechmea*. Note yellowing of basal leaves.

78. Rhizoctonia aerial blight on *Aechmea fasciata*.

79. Slime mold on bromeliad.

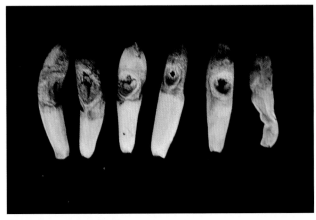

80. Anthracnose on *Sedum*.

81. Botrytis blight on *Senecio*.

82. Damage to *Schlumbergera* caused by cactus cyst nematode.

83. Cercospora leaf spot on *Crassula*.

84. Dichotomophthora leaf spot on *Opuntia*.

85. Erwinia blight on *Schlumbergera*.

86. Erwinia blight on *Haworthia*.

87. Ethylene damage to *Crassula*.

88. Fusarium cladophyll rot on *Schlumbergera*.

89. Fusarium stem rot on *Ferocactus*.

90. Fusarium crown rot on *Faucaria.*

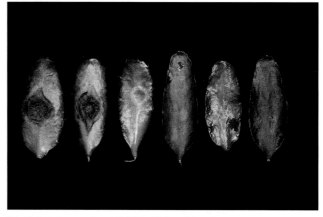

91. Helminthosporium blight in *Rhipsalidopsis.*

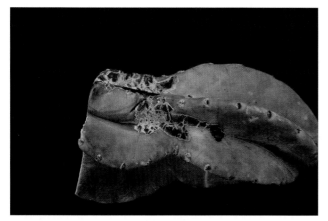

92. Helminthosporium stem rot on *Cereus.*

93. Helminthosporium stem rot on *Opuntia.*

94. Phytophthora stem rot on *Schlumbergera.*

95. Pythium root rot on *Rhipsalidopsis.*

96. Aspergillus corm rot on *Caladium*.

97. Botrytis blight on *Caladium*.

98. Cold damage on *Caladium* leaves. (Courtesy G. W. Simone)

99. Dasheen mosaic on *Caladium*. (Courtesy G. W. Simone)

100. Stunting of *Caladium* (plant on right) caused by Fusarium corm rot.

101. Fusarium corm rot on *Caladium* corms.

102. Rhizoctonia corm rot on sprouting *Caladium* corms.

103. Stunting of *Caladium* (plants on left) caused by root-knot nematodes.

104. *Caladium* with southern blight (note sclerotia in center). (Courtesy J. F. Knauss)

105. *Caladium* with sunscald damage. (Courtesy J. F. Knauss)

106. Xanthomonas blight on *Caladium*.

107. Alternaria leaf spot on *Calathea insignis*.

108. Stunting of *Calathea insignis* caused by burrowing nematodes (plant on far right is healthy).

109. Cucumber mosaic on *Calathea*.

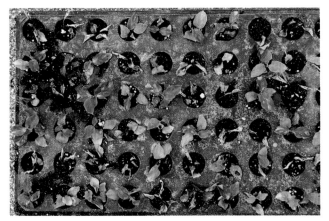

110. Desiccation damage on tissue-cultured *Calathea*.

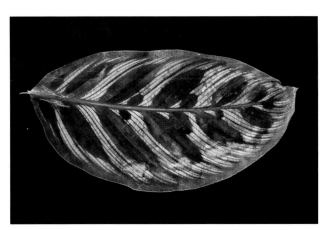

111. Fluoride toxicity on *Calathea makoyana*.

112. Damage to *Calathea* caused by foliar nematodes. (Courtesy W. J. Nishijima)

113. Fusarium wilt on *Calathea*. (Courtesy J. M. F. Yuen)

114. Helminthosporium leaf spot on *Calathea*.

115. Potassium deficiency on *Calathea makoyana*.

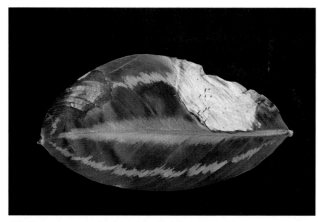

116. Pseudomonas leaf spot on *Calathea*.

117. Pseudomonas blight on *Calathea roseo-lineata*.

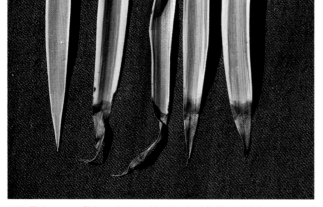

118. Tipburn on *Chlorophytum*. Healthy leaf (left) and tipburn caused by fluoride toxicity (next two) and boron toxicity (two on right).

119. Anthracnose on *Cissus antarctica*.

120. Window-pane effect on *Cissus antarctica* caused by anthracnose.

121. Bendiocarb damage on *Cissus rhombifolia*. (Courtesy J. M. F. Yuen)

122. Botrytis blight on *Cissus* leaves.

123. Downy mildew on *Cissus*.

124. Powdery mildew on *Cissus*. (Courtesy J. M. F. Yuen)

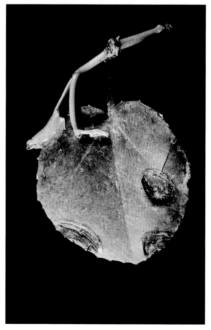

125. Rhizoctonia leaf spot on *Cissus*. (Courtesy Florida Division of Plant Industry)

126. Anthracnose on *Codiaeum*.

127. Botrytis blight on *Codiaeum* tip cuttings.

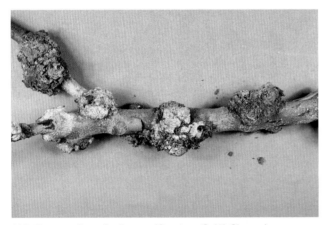

128. Crown gall on *Codiaeum*. (Courtesy G. W. Simone)

129. Fusarium stem rot on *Codiaeum* cuttings.

130. Fusarium root rot on *Codiaeum* cuttings. (Courtesy G. W. Simone)

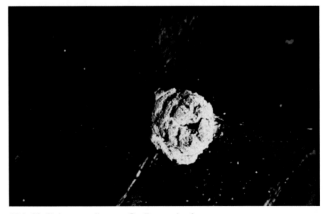

131. Kutilakesa gall on a *Codiaeum* leaf.

132. Stunting of *Codiaeum* caused by fertilizer deficiency (plant on the left received appropriate level of fertilizer).

133. Xanthomonas leaf spot on *Codiaeum*.

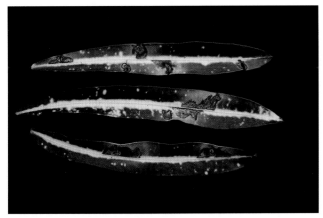

134. Xanthomonas leaf spot on *Codiaeum*. (Courtesy J. M. F. Yuen)

135. Botrytis blight on *Coleus*.

136. Chimera on *Coleus* (purple coloration is not normal for this cultivar).

137. Corynespora leaf spot on *Coleus*.

117

138. Pseudomonas leaf spot on *Coleus*. (Courtesy Florida Division of Plant Industry)

139. Damage to *Coleus* caused by root-knot nematodes. (Courtesy G. W. Simone)

140. Sunscald on *Coleus* leaves (dark spots on yellow tissue).

141. Corynespora leaf spot on *Columnea*.

142. Rhizoctonia aerial blight on *Columnea*. (Courtesy R. W. Henley)

143. Anthracnose on *Cordyline*.

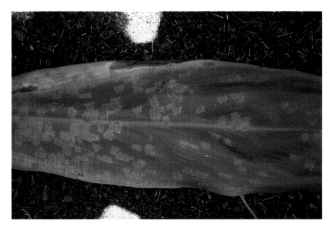

144. Cercospora leaf spot on *Cordyline*.

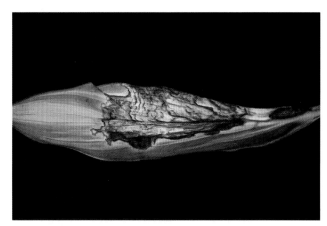

145. Erwinia leaf blight on *Cordyline*.

146. Erwinia stem rot on *Cordyline* cuttings.

147. Erwinia root rot on *Cordyline* cuttings.

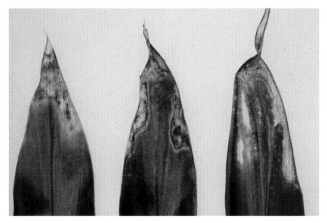

148. Fluoride toxicity on *Cordyline*.

149. Fusarium root rot on *Cordyline*.

150. Fusarium stem and root rot on *Cordyline*.

151. Phyllosticta leaf spot on *Cordyline*.

152. Southern blight on *Cordyline* cuttings.

153. Xanthomonas blight on *Cordyline*.

154. Anthracnose on *Dieffenbachia* 'Camille.'

155. Cold damage on *Dieffenbachia* 'Camille.' (Courtesy J. M. F. Yuen)

156. Copper toxicity on *Dieffenbachia* 'Perfection.'

157. Stunting of *Dieffenbachia* 'Perfection' (left) caused by dasheen mosaic virus.

158. Ring spots on *Dieffenbachia* 'Rudolph Roehrs' caused by dasheen mosaic virus.

159. Mosaic on *Dieffenbachia* 'Perfection' caused by dasheen mosaic virus.

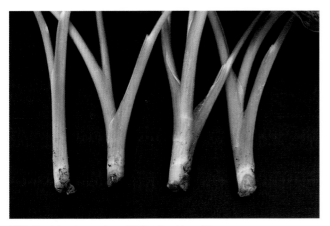

160. Erwinia stem rot on *Dieffenbachia* cuttings.

161. Erwinia leaf spot on *Dieffenbachia* 'Perfection.'

162. Fusarium cutting rot on *Dieffenbachia.*

163. Fusarium stem rot on tissue-cultured plantlets of *Dieffenbachia.*

164. Leptosphaeria leaf spot on *Dieffenbachia maculata.*

165. Myrothecium leaf spot on *Dieffenbachia* 'Camille.'

166. Phytophthora aerial blight on *Dieffenbachia amoena.*

167. Pythium root rot on *Dieffenbachia* 'Tropic Snow.'

168. Sunscald on *Dieffenbachia*.

169. Xanthomonas blight on *Dieffenbachia* 'Camille.'

170. Anthracnose on Venus's fly-trap. (Courtesy G. W. Simone)

171. Pythium root rot on Venus's fly-trap. (Courtesy Florida Foliage Association)

172. Anthracnose on *Dracaena sanderana*.

173. Botrytis blight on *Dracaena godseffiana* cuttings.

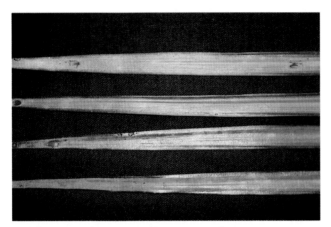

174. Cercospora leaf spot on *Dracaena marginata.*

175. Cold damage on *Dracaena massangeana.*

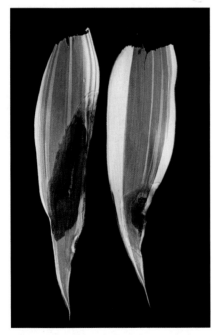

176. Leaf spot on *Dracaena sanderana* caused by *Erwinia carotovora.*

177. Erwinia blight on *Dracaena* 'Warneckii.'

178. Leaf spot on *Dracaena sanderana* caused by *Erwinia herbicola*.

179. Flecking disorder on *Dracaena marginata*.

180. Fluoride toxicity on *Dracaena* 'Warneckii.'

181. Fluoride toxicity on *Dracaena massangeana*.

182. Fusarium stem rot on *Dracaena massangeana.*

183. Fusarium leaf spot on *Dracaena marginata.*

184. Fusarium cutting rot on *Dracaena marginata.*

185. Notching on *Dracaena* 'Warneckii.'

186. Phyllosticta leaf spot on *Dracaena sanderana.*

187. Pseudomonas leaf spot on *Dracaena sanderana.*

188. Yellowing of new leaves on *Dracaena deremensis* caused by Pythium root rot.

189. Anthracnose on pothos.

190. Bird's nest fungi on undersides of pothos leaves. (Courtesy G. W. Simone)

191. Chilling injury on Marble Queen pothos.

192. Erwinia petiole rot on Golden pothos.

193. Manganese toxicity on pothos.

194. Nitrogen toxicity on Golden pothos.

195. Potassium deficiency on pothos (right).

196. Pseudomonas leaf spot on pothos.

197. Pythium root rot on pothos. (Courtesy J. M. F. Yuen)

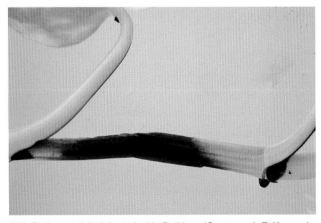

198. Pothos petiole infected with *Pythium*. (Courtesy J. F. Knauss)

199. Rhizoctonia foot rot on pothos. Stems, petioles, and leaves are affected.

200. Southern blight on Golden pothos. (Courtesy J. M. F. Yuen)

201. Xanthomonas blight on satin pothos.

202. Fusarium blight on *Episcia*.

203. Rhizoctonia aerial blight on *Episcia*.

204. Erwinia soft rot on *Euphorbia milii*.

205. Fusarium stem rot on *Euphorbia*.

206. Myrothecium leaf spot on *Euphorbia*.

207. Phomopsis stem blight on *Euphorbia trigona*.

208. Phytophthora stem rot on *Euphorbia ingens*.

209. Rhizoctonia stem rot on *Euphorbia milii*.

210. Rhizopus blight on *Euphorbia milii*.

130

211. Sunscald on *Euphorbia*. (Courtesy G. W. Simone)

212. Xanthomonas leaf spot on *Euphorbia*.

213. Anthracnose on *Ficus elastica*.

214. Bendiocarb toxicity on *Ficus elastica*.

215. Boron toxicity on *Ficus benjamina*.

216. Botrytis on cuttings of *Ficus elastica*.

217. Bromine toxicity on *Ficus benjamina*. (Courtesy J. M. F. Yuen)

218. Cercospora leaf spot on *Ficus elastica*.

219. Corynespora leaf spot on variegated *Ficus benjamina*.

220. Crown gall on *Ficus benjamina*. (Courtesy H. Bouzar)

221. Foliar nematode damage on *Ficus elastica*. (Courtesy J. M. F. Yuen)

222. Frost injury on *Ficus elastica*.

223. Helminthosporium leaf spot on *Ficus lyrata*.

224. Myrothecium leaf spot on *Ficus benjamina*.

225. Myrothecium leaf spot on *Ficus elastica*.

226. Myrothecium stem rot on *Ficus pumila*.

227. Phomopsis canker on *Ficus benjamina*.

228. Phomopsis streaking in wood of *Ficus benjamina*.

229. Phomopsis blight on *Ficus benjamina*.

230. Pseudomonas leaf blight on *Ficus lyrata*. (Courtesy J. M. F. Yuen)

231. Rhizoctonia aerial blight on *Ficus benjamina*.

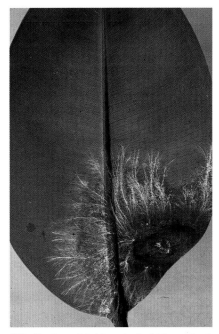

232. Southern blight on leaf of *Ficus elastica*. (Courtesy G. W. Simone)

233. Southern blight on stems of *Ficus pumila*. Note sclerotia. (Courtesy J. M. F. Yuen)

234. Xanthomonas blight on *Ficus benjamina*.

235. Distortion caused by Biden's mottle virus on *Fittonia*.

236. Distortion and stunting of *Fittonia* leaves caused by Biden's mottle virus. Leaf on far left is normal.

237. Chilling injury on *Fittonia*.

238. Rhizoctonia aerial blight on *Fittonia*.

239. Xanthomonas blight on *Fittonia*. (Courtesy J. M. F. Yuen)

136

240. Anthracnose on *Hedera*.

241. Botrytis blight on *Hedera*.

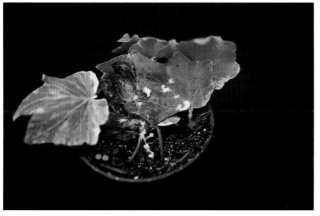

242. Sclerotinia blight on *Hedera*.

243. Bromine toxicity on *Hedera*.

244. Chlorothalonil toxicity on *Hedera*.

245. Fertilizer burn on *Hedera*. (Courtesy J. M. F. Yuen)

246. Fusarium leaf spot on *Hedera*.

247. Fusarium and Pythium root rot on *Hedera*.

248. Pythium root rot on *Hedera*.

249. Rhizoctonia aerial blight on *Hedera*.

250. Streptomycin toxicity on *Hedera*.

251. Xanthomonas leaf spot on *Hedera*.

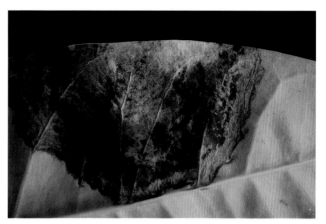

252. Botrytis blight on *Hoya*.

253. Myrothecium leaf spot on *Hoya*.

254. Calonectria stem rot on *Leea*.

255. Phytophthora blight on *Leea*.

256. Phytophthora leaf spot on *Leea*.

257. Xanthomonas leaf spot on *Leea*.

258. Benomyl toxicity on *Maranta erythroneura* (leaf on far right is normal color).

259. Bleach toxicity on *Maranta leuconeura*.

260. Cucumber mosaic on *Maranta leuconeura*.

261. Cucumber mosaic on *Maranta erythroneura* (leaf on the left is healthy).

262. Helminthosporium leaf spot on *Maranta*.

263. Pythium root rot in ground beds of *Maranta*.

264. Edema on *Pilea cadierei*.

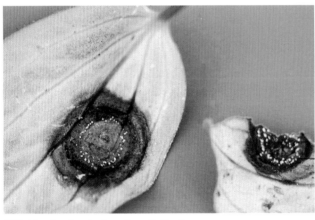
265. Myrothecium leaf spot on *Pellionia*.

266. Pythium root rot on *Pilea cadierei*.

267. Rhizoctonia aerial blight on *Pilea serpyllacea*.

268. Xanthomonas blight on *Pilea cadierei*.

141

269. Xanthomonas blight on *Pellionia*.

270. Anthracnose on *Peperomia*. (Courtesy G. W. Simone)

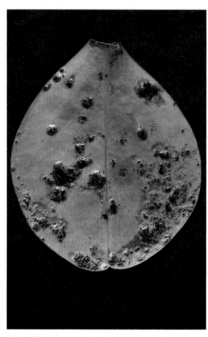

271. Cercospora leaf spot on *Peperomia obtusifolia*.

272. Myrothecium leaf spot on *Peperomia*.

273. Phyllosticta leaf spot on *Peperomia*.

274. Phytophthora stem rot on *Peperomia obtusifolia*. (Courtesy J. M. F. Yuen)

142

275. Stunting of *Peperomia obtusifolia* (left) caused by Pythium root rot.

276. Rhizoctonia leaf spot on *Peperomia obtusifolia.*

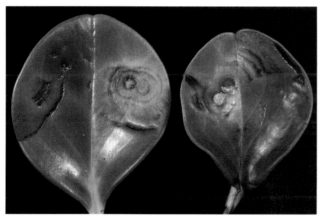

277. Peperomia ring spot virus on *Peperomia obtusifolia.* (Courtesy G. W. Simone)

278. Southern blight on *Peperomia obtusifolia.* (Courtesy G. W. Simone)

279. Bird's nest fungus on *Philodendron.*

280. Bird's nest fern fungus.

143

281. Botrytis blight on *Philodendron scandens* subsp. *oxycardium*. (Courtesy G. W. Simone)

282. Dactylaria leaf spot on *Philodendron scandens* subsp. *oxycardium*.

283. Dasheen mosaic virus infection of *Philodendron*.

284. Erwinia blight on *Philodendron selloum*.

285. Erwinia leaf spot on *Philodendron*.

286. Heater damage to *Philodendron scandens* subsp. *oxycardium*.

144

287. Phytophthora leaf spot on *Philodendron scandens* subsp. *oxycardium*.

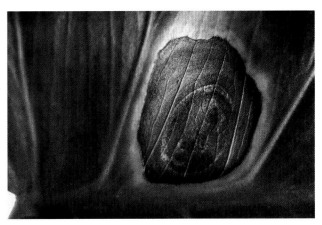

288. Pseudomonas leaf spot on *Philodendron*.

289. Pythium root rot on *Philodendron scandens* subsp. *oxycardium*.

290. Red-edge disease on *Philodendron scandens* subsp. *oxycardium*.

291. Rhizoctonia leaf spot on *Philodendron scandens* subsp. *oxycardium*. (Courtesy J. M. F. Yuen)

292. Southern blight on *Philodendron scandens* subsp. *oxycardium*.

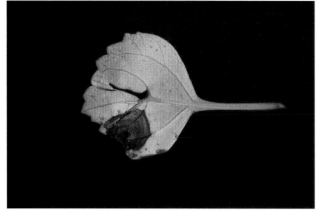

293. Myrothecium leaf spot on Swedish ivy.

294. Pseudomonas leaf spot on Swedish ivy. (Courtesy G. W. Simone)

295. Rhizoctonia aerial blight on Swedish ivy.

296. Botrytis blight on *Nephrolepis*. (Courtesy G. W. Simone)

297. Drought damage to *Nephrolepis*. Plant on left is healthy.

298. Foliar nematode damage on *Asplenium*.

299. Herbicide damage on *Nephrolepis*.

300. Myrothecium leaf spot on *Dryopteris*.

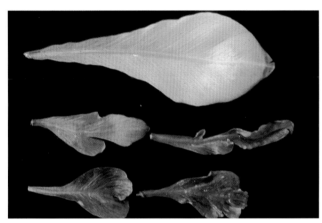
301. Nitrogen toxicity on *Asplenium* (top leaf is healthy).

302. Pseudomonas blight on *Asplenium*.

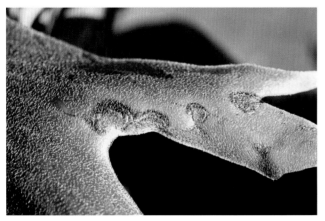
303. Pseudomonas leaf spot on *Platycerium*.

304. Pseudomonas leaf spot on *Pteris*.

305. Rhizoctonia aerial blight on *Asparagus*.

306. Rhizoctonia aerial blight on *Nephrolepis*.

307. Rhizoctonia leaf spot on *Cyrtomium*.

308. Alternaria leaf spot on *Fatshedera*.

309. Alternaria leaf spot on *Polyscias fruticosa*.

310. Anthracnose on *Polyscias balfouriana*. (Courtesy J. M. F. Yuen)

311. Crown gall on roots of *Polyscias fruticosa*. (Courtesy G. W. Simone)

312. Fusarium stem rot on *Fatsia japonica*.

313. Fusarium stem canker on *Fatsia japonica*.

314. Phytophthora aerial blight on *Fatsia japonica*. (Courtesy Florida Division of Plant Industry)

315. Pseudomonas leaf spot on *Fatsia japonica*.

316. Pseudomonas leaf spot on *Fatshedera*.

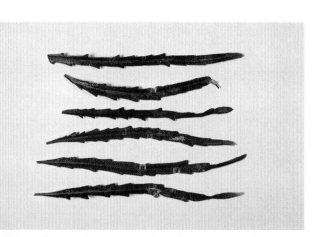

317. Pseudomonas leaf spot on *Dizygotheca elegantissima*.

318. Root-knot nematode galls on roots of *Polyscias*. (Courtesy R. W. Henley)

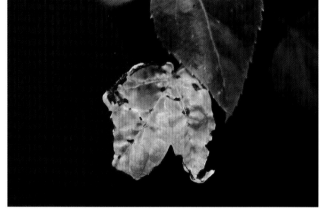

319. Xanthomonas blight on *Fatsia japonica*.

320. Xanthomonas leaf spot on *Polyscias*.

321. Xanthomonas leaf spot on *Dizygotheca elegantissma*.

322. Corynespora leaf spot on *Radermachera*.

323. Damping-off of *Radermachera* seedlings.

324. Fusarium stem rot on *Radermachera*.

325. Rhizoctonia stem rot on *Radermachera*.

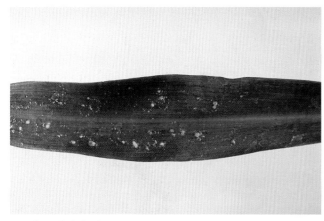

326. Curvularia leaf spot on *Rhoeo*.

327. Tobacco mosaic virus infection of *Rhoeo*.

328. Botrytis blight on African violet flowers.

329. Botrytis blight on African violet leaves.

330. Chimera (genetic variation) on African violet.

331. Cold water damage on African violet.

332. Corynespora leaf spot on African violet.

333. Erwinia blight of African violet.

334. Ethylene damage to flowers of African violet.

335. Ethylene damage to leaves of African violet.

336. Foliar nematode damage on African violet.

337. Phytophthora stem rot on African violet.

338. Phytophthora stem rot on African violet plantlets.

339. Powdery mildew on African violet.

 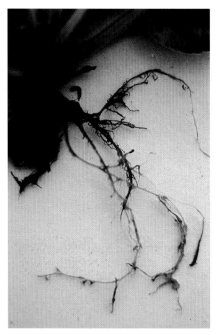

340. Pythium root rot on African violet. (Courtesy G. W. Simone)

341. Root-knot nematode damage on African violet.

342. Tomato spotted wilt on African violet.

343. Aspergillus rhizome rot on *Sansevieria*.

344. Cold damage on *Sansevieria*. (Courtesy G. W. Simone)

345. Fusarium leaf spot on bird's nest *Sansevieria*.

346. Soft rot on *Sansevieria*.

347. Acephate toxicity on schefflera.

348. Alternaria leaf spot on schefflera.

349. Alternaria leaf spot on dwarf schefflera.

350. Ammonium toxicity on schefflera.

351. Bendiocarb toxicity on schefflera. (Courtesy J. M. F. Yuen)

352. Chemical phytotoxicity on schefflera. (Courtesy G. W. Simone)

353. Dodder on schefflera. (Courtesy R. W. Henley)

354. Edema on schefflera.

355. Frost damage on schefflera.

356. Damage from spider mite feeding on schefflera grown indoors.

357. Phytophthora leaf spot on schefflera.

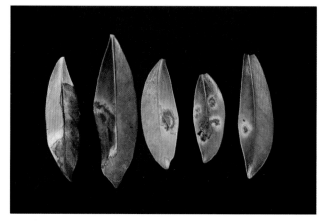

358. Pseudomonas leaf spot on dwarf schefflera.

359. Pythium root rot on schefflera.

360. Rhizoctonia damping-off on schefflera seedlings.

361. Xanthomonas leaf spot on dwarf schefflera.

362. Xanthomonas leaf spot on schefflera.

363. Algae growth on *Spathiphyllum* leaves.

364. Benomyl toxicity on *Spathiphyllum*.

365. Chimera on *Spathiphyllum*.

366. Chlorosis on *Spathiphyllum* caused by fertilizer deficiency (plant on the far left is healthy).

367. Cylindrocladium leaf spot on *Spathiphyllum*. (Courtesy J. M. F. Yuen)

368. Cylindrocladium petiole damping-off on *Spathiphyllum* plugs.

369. Cylindrocladium petiole rot on *Spathiphyllum*.

370. Cylindrocladium root and petiole rot on *Spathiphyllum*.

371. Dasheen mosaic virus infection on *Spathiphyllum*.

372. Erwinia spadix rot on *Spathiphyllum* flower. (Courtesy J. M. F. Yuen)

373. Erwinia stem rot on *Spathiphyllum*.

374. Fertilizer burn on *Spathiphyllum*.

375. Magnesium deficiency on lower leaves of *Spathiphyllum*.

376. Myrothecium petiole rot on *Spathiphyllum* plug.

377. Myrothecium leaf spot on *Spathiphyllum*. (Courtesy J. M. F. Yuen)

378. Phytophthora aerial blight on *Spathiphyllum*.

379. Phytophthora aerial blight on *Spathiphyllum*.

380. Southern blight on leaves of *Spathiphyllum*. (Courtesy G. W. Simone)

381. Acremonium leaf spot on *Syngonium*.

382. Ceratocystis cane rot on *Syngonium.*

383. Cold water damage on *Syngonium.*

384. Copper toxicity on *Syngonium.*

385. Erwinia leaf spot on *Syngonium.*

386. Myrothecium leaf spot on *Syngonium.*

387. Myrothecium petiole rot on *Syngonium.*

388. Rhizoctonia leaf spot on *Syngonium*. (Courtesy G. W. Simone)

389. Xanthomonas blight on *Syngonium*.

390. Xanthomonas leaf spot (blight) on *Syngonium*.

391. Xanthomonas leaf spot (blight) on *Syngonium*.

392. Anthracnose on piggyback plant. (Courtesy L. Pierce)

393. Pythium root rot on *Tradescantia*. (Courtesy G. W. Simone)

394. Distortion of *Tradescantia* caused by Tradescantia virus.

395. Myrothecium petiole rot on *Xanthosoma*.

396. Myrothecium leaf spot on *Xanthosoma*.

397. Xanthomonas blight on *Xanthosoma*.

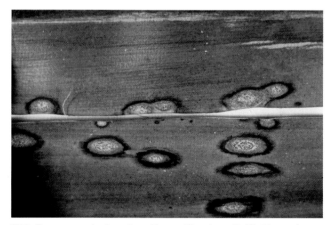

398. Cercospora leaf spot on *Yucca*. (Courtesy G. W. Simone)

399. Coniothyrium (*Microsphaeropsis*) leaf spot on *Yucca*.

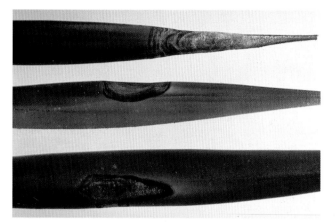

400. Cytosporina leaf spot on *Yucca*.

401. Fusarium cutting rot on *Yucca*.

402. Fusarium fruiting bodies on *Yucca*.

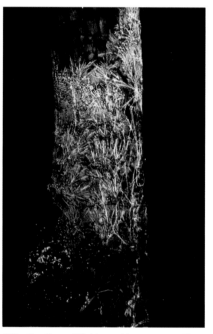

403. Southern blight on *Yucca* cane. Note sclerotia.

Index

Acephate phytotoxicity, 75; Fig. 347
Acremonium crotocinigenum, 81
Acremonium leaf spot on *Syngonium*, 81; Fig. 381
Agrobacterium tumefaciens, 33, 51, 69
algae on *Spathiphyllum*, 78; Fig. 363
Aloe, 23
Alternaria
　alternata, 29
　panax, 68, 75
Alternaria leaf spot
　on *Calathea*, 29; Fig. 107
　on *Polyscias* and related plants, 68; Figs. 308
　　and 309
　on schefflera, 75; Figs. 348 and 349
ammonium toxicity, 76; Fig. 350
anthracnose
　on *Aglaonema*, 8; Fig. 7
　on *Anthurium*, 11; Fig. 24
　on *Araucaria*, 15; Fig. 43
　on cissus, 32; Figs. 119 and 120
　on croton, 33; Fig. 126
　on *Dieffenbachia*, 38; Fig. 154
　on *Dionaea*, 41; Fig. 170
　on dracaena, 41; Fig. 172
　on English ivy, 54; Fig. 240
　on *Ficus*, 49; Fig. 213
　on palms, 16; Fig. 44
　on *Peperomia*, 60; Fig. 270
　on piggyback plant, 83; Fig. 392
　on *Polyscias* and related plants, 69; Fig. 310
　on pothos, 44; Fig. 189
　on succulents, 24; Fig. 80
　on ti plant, 36; Fig. 143
Aphelenchoides
　besseyi, 30, 51
　fragariae, 66
　ritzemabosi, 73
Aspergillus niger, 27, 74
Aspergillus rot
　on *Caladium*, 27; Fig. 96
　on snake plant, 74; Fig. 343
Asplenium, 66

bendiocarb phytotoxicity
　on *Aphelandra*, 13; Fig. 34
　on cissus, 32; Fig. 121
　on *Ficus*, 50; Fig. 214

on schefflera, 76; Fig. 351
benomyl toxicity
　on prayer plant, 58; Fig. 258
　on *Spathiphyllum*, 78; Fig. 364
bent tip on *Aglaonema*, 8; Figs. 8 and 9
Bidens mottle on *Fittonia*, 53; Figs. 235 and 236
Bidens mottle virus, 53
Bipolaris spp., 17, 51
bird's nest fungi
　on philodendron, 62; Figs. 279 and 280
　on pothos, 45; Fig. 190
black cane rot, 81; Fig. 382
bleach toxicity on prayer plant, 58; Fig. 259
boron deficiency on palms, 16; Fig. 45
boron toxicity
　on *Aglaonema*, 8; Fig. 10
　on *Chlorophytum*, 31; Fig. 118
　on *Ficus*, 50; Fig. 215
Botrytis blight
　on *Aeschynanthus*, 7; Fig. 1
　on African violet, 71; Figs. 328 and 329
　on begonia, 21; Fig. 64
　on *Caladium*, 27; Fig. 97
　on *Cissus*, 32; Fig. 122
　on *Coleus*, 34; Fig. 135
　on croton, 33; Fig. 127
　on dracaena, 42; Fig. 173
　on English ivy, 55
　on ferns, 66; Fig. 296
　on *Ficus*, 50; Fig. 216
　on philodendron, 63; Fig. 281
　on *Senecio*, 24; Fig. 81
　on wax plant, 57; Fig. 252
Botrytis cinerea, 7, 21, 24, 27, 32, 33, 34, 42, 50,
　　57, 63, 66, 71
bromine toxicity
　on English ivy, 55; Fig. 243
　on *Ficus*, 50; Fig. 217
brown leaf spot
　on *Dieffenbachia*, 39; Fig. 164
　on yucca, 85; Fig. 399
burrowing nematode on *Calathea*, 29; Fig. 108

Cactodera cacti, 24
Cactus cyst nematode on cacti, 24; Fig. 82
Calonectria
　colhounii, 16